Information Operations Planning

For a complete listing of the *Artech House Information Warfare Series,*
turn to the back of this book.

Information Operations Planning

Patrick D. Allen

ARTECH HOUSE
BOSTON | LONDON
artechhouse.com

Library of Congress Cataloging-in-Publication Data
A catalog record for this book is available from the U.S. Library of Congress.

British Library Cataloguing in Publication Data
A catalogue record for this book is available from the British Library.

Cover design by Yekatarina Ratner

© 2007 Patrick D. Allen

Published by Artech House
685 Canton Street
Norwood MA 02062

All rights reserved. Printed and bound in the United States of America. No part of this book may be reproduced or utilized in any form or by any means, electronic or mechanical, including photocopying, recording, or by any information storage and retrieval system, without permission in writing from the publisher.

All terms mentioned in this book that are known to be trademarks or service marks have been appropriately capitalized. Artech House cannot attest to the accuracy of this information. Use of a term in this book should not be regarded as affecting the validity of any trademark or service mark.

ISBN 10: 1-58053-517-8
ISBN 13: 978-1-58053-517-5

10 9 8 7 6 5 4 3 2 1

*To my wife, Moira,
not only for her support throughout this effort,
but also for being a darn good editor!*

Contents

	Foreword	**xv**
	Part I: Information Operations Planning Context	**1**
1	**Information Operations and Its Role in Security and Conflict**	**3**
1.1	The Purposes and Principles of Information Operations: What Do You Think You Know?	3
1.2	Examples of IO Implementations in Recent Conflicts	5
1.2.1	Operation Desert Storm (ODS)	6
1.2.2	Serb IO in the Kosovo Campaign	6
1.2.3	Allied IO in the Kosovo Campaign	7
1.2.4	Internet Conflicts	8
1.2.5	Allied IO in Operation Iraqi Freedom	9
1.2.6	IO After the Liberation of Iraq	10
1.2.7	Cultural Messages via the Media	12
1.3	What IO Is Not	14
1.4	The Role of IO in Effects-Based Operations (EBO)	18
	References	21
2	**The Increasing Importance of IO in Military Operations**	**25**
2.1	Early Examples of Information Operations	25

2.2	Military Conflict in the Information Age	27
2.3	Advantage, Dependence, and Vulnerability in Information Age Societies	29
	References	34

3 Characteristics of IO Planning — 37

3.1	Feature 1: The Large IO Planning Space	38
3.2	Feature 2: The Time and Space Dimensions of IO	40
3.2.1	Temporal Scope	40
3.2.2	Physical Scope	42
3.3	Feature 3: Direct and Indirect Effects in IO	43
3.4	Feature 4: The Role of Feedback in IO	44
3.4.1	Does He Know That I Know … ?	46
3.5	Feature 5: The Complexity of IO	47
3.5.1	Organizing to Handle Complexity	48
3.6	Feature 6: The Information Requirements of IO	49
3.6.1	Information Readiness	50
3.7	Feature 7: IO Classification and Protection Issues	50
	References	52

Part II: The IO Planning Process and Tools — 55

4 Categories of IO Planning — 57

4.1	The Evolving U.S. Doctrine of IO	57
4.2	Three Useful Categories of IO	59
4.3	Types of IO Operations Planning	62
4.3.1	Deliberate Planning	62
4.3.2	Crisis Action Planning	63
4.3.3	Execution Planning	64
4.3.4	Persistent Planning	65
4.3.5	Execution Monitoring and Decision Support	66
4.3.6	Postcrisis Assessment and Feedback	67
4.4	Organizing for IO Operational Planning	68

4.5	IO Acquisition Planning	69
	References	71

5	**Planning Methodologies and Technologies for Attack Missions in IO**	**73**
5.1	Defining the Space of Options	74
5.2	Quantifying the Political Constraints	80
5.3	Cause-and-Effect Networks (CAENs)	87
5.4	Blue-Versus-Red COA Comparison	92
5.4.1	Comparing Blue and Red COAs	94
5.5	Employment Options and the Blue Capabilities Matrix	97
5.6	Scalability, Reachback, and Requests for Analysis	101
5.7	Multilevel Security and Distributed Plans	104
	References	107

6	**Planning Methodologies and Technologies for Influence Missions in IO**	**109**
6.1	The Influence Operations Planning Context	110
6.2	Categorizing Message Purposes	112
6.3	The Idea Battlespace	114
6.3.1	Messages Interacting with Groups	116
6.3.2	Messages Interacting with Messages	117
6.3.3	Groups Interacting with Groups	118
6.4	Getting and Maintaining Attention	119
6.5	Change Methods and Submethods	120
6.5.1	Creating a New Message	121
6.5.2	Modifying an Existing Message	122
6.5.3	Changing the Level of Attention	122
6.5.4	Co-Opting an Existing Message	122
6.5.5	Subverting a Message	123
6.5.6	Constraints on Public Affairs Personnel in the Idea Battlespace	126

6.6	Delivery Mechanisms	127
6.7	Measuring Influence Success	127
6.8	Entering Influence Operations in the CAEN	129
	References	131

7 Planning Methodologies and Technologies for Defense Missions in IO — 133

7.1	How Defense Differs from Attack and Influence	134
7.2	Military Deception Planning	135
7.3	Operations Security Planning	140
7.3.1	Additional Issues in MILDEC and OPSEC Planning	144
7.4	Computer Network Defense Planning	146
7.4.1	Network Defense Planning Principles	147
	References	153

8 Monitoring and Replanning During Execution — 155

8.1	Monitoring Plan Progress	156
8.1.1	Planning for Monitoring	157
8.1.2	Monitoring Plan Execution	159
8.2	Monitoring Branches and Sequels	161
8.2.1	Sequels	162
8.2.2	Branches	163
8.2.3	Monitoring Decision Points	165
8.3	Execution Monitoring for Attack	166
8.4	Execution Monitoring for Influence	169
8.5	Execution Monitoring for Defense	171
8.5.1	Monitoring Military Deception and OPSEC Plans	172
8.5.2	Monitoring Computer Network Defense	174
8.6	When to Replan—or Not Replan	175
8.7	Feedback Process to Refine Planning Factors	177
	References	179

9	**Planning Tools and Technologies**	**181**
9.1	Semiautomating the Planning Process	182
9.1.1	Why Semiautomated Planning Is Needed	182
9.1.2	Semiautomated Planning Versus Automated Planning	183
9.1.3	Human and Machine Strengths and Weaknesses	184
9.1.4	The Need for Visualization	184
9.1.5	The Need for Evidential Reasoning Tools	185
9.1.6	Planning Process Workflows	186
9.2	Semiautomated Planning Tools	189
9.2.1	Current Planning Tools	190
9.2.2	The Echelons of Focus of Each Planning Tool	190
9.2.3	Current IO Planning Tools	193
9.3	COA Planning Technologies	195
9.3.1	The COA Development Process	195
9.3.2	Types of COAs	196
9.3.3	The Course of Action Support Tool in IWPC	197
9.3.4	Other COA Development Tools for EBP	200
9.4	Effects Prediction Technologies	204
9.4.1	Benefits of Prediction Technologies	204
9.4.2	The Tradeoffs of Modeling and Simulation Tools	205
9.4.3	Effects Prediction Tools for Planning	207
9.5	Adversarial and Nonadversarial Reasoning Technologies	208
9.5.1	Adversarial Reasoning	209
9.5.2	Nonadversarial Reasoning: Shifting Allegiances in Combat Situations	212
9.5.3	Nonadversarial Reasoning: Shifting Allegiances in Noncombat Situations	215
	References	219
10	**Planning for IO Logistics, Training, and Acquisition**	**223**
10.1	The Logistics of IO	224
10.1.1	Information Requirements for IO Planning	224
10.1.2	Information Suppliers	227
10.1.3	Information Readiness	228

10.2	IO Transformation and Acquisition Planning	230
10.2.1	DOTMLPF	230
10.2.2	Planning Methodologies for IO Acquisition	233
10.2.3	Acquiring Civil Cyber-Defense	236
10.3	IO in Training, Experimentation, and Testing	237
10.3.1	IO in Training	237
10.3.2	How to Include CNO in Exercises and Experiments	239
10.3.3	IO in Experimentation	241
10.3.4	IO in Testing	242
10.4	Legal Issues of IO	246
10.4.1	What Is a Hostile Act?	246
10.4.2	What Is Legally Allowed in IO?	247
10.4.3	How to Help Planners Know Up-Front What Is Allowed	250
10.4.4	Title 10 and Title 50 Legal Issues	251
	References	252

Part III: The Future of IO Planning — 255

11 The Future of Information Conflict — 257

11.1	Feature 1: Everyone Has a Message	258
11.2	Feature 2: Every Idea Can Rapidly Be Known Globally	260
11.3	Feature 3: Everyone in the World Is the Potential Audience for Every Message We Create	261
11.4	Feature 4: Every Group Has Its Own Spin on Every Other Group's Message	262
11.5	Feature 5: Most Messages Are *About* a Third Group, Rather Than a Direct Communication Between Two Groups	264
11.6	Feature 6: Initiative Matters in the Idea Battlespace	264
11.7	Feature 7: Accusations Are an Easy Way to Gain the Initiative	266
11.8	Feature 8: Conflicts in the Idea Battlespace Will Often Spill Over into the Physical World	267

11.9	Feature 9: Conflicts in the Idea Battlespace Cannot Be Won in the Physical Space	268
11.10	Feature 10: Allies Are Important in Competitions Within the Idea Battlespace	269
11.11	Feature 11: Neither Adversaries Nor Allies Are Necessarily Permanent in the Idea Battlespace	270
11.12	Feature 12: Some Groups Will Cheat in the Idea Battlespace and Thereby Gain an Advantage	272
11.13	Feature 13: Multiple Dimensions of Power Are Essential to the Long-Term Success in the Idea Battlespace	274
11.14	Feature 14: There Is No Rear Area or Sanctuary in the Idea Battlespace	275
11.15	Feature 15: Conflict in the Idea Battlespace Is Persistent, Not Episodic	278
	References	279
12	**Planning for Success in a New Age**	**283**
12.1	Revolutions in Military and Political Thought	284
12.1.1	Recognizing That We Are in a New Age	284
12.1.2	The Information Age: What's New and What's Changed	285
12.1.3	A Problem and an Opportunity	287
12.2	The Information Sphere as a Domain	289
12.2.1	Definitions	290
12.2.2	Features of a Domain	294
12.2.3	Differences of the Information Sphere's Domain	297
12.2.4	Benefits of the DoD's Treating the Information Sphere as a Domain	300
12.2.5	How New Domains Are Formed	302
	References	305
	Appendix: IO Attack Desired Effects Definitions	**307**
	List of Leading Quote References	**313**
	About the Author	**315**
	Index	**317**

Foreword

I first met Pat Allen over five years ago when he was briefing newly-developed tools to help military Information Operations (IO) planners develop complex courses of action that integrated diplomatic, information, military and economic actions. In those days, his work was tightly held and very little was published about his planning research. Now he is able to publish a significant contribution to the IO literature that captures the state-of-the-art in campaign planning methods, tools and technologies.

Campaign planning is in the process of significant change. Far beyond traditional military planning, the process increasingly requires the coordination of all elements of power and persuasion to achieve precision effects on targets across a wide spectrum of political-social, economic, military, information and infrastructure systems. Planning teams now encompass participants from many agencies of government: diplomats, economists, urban planners, and other domain experts join with military commanders. Using collaborative planning tools, these teams describe plan concepts, explore the complex interactions between planned action sequences and their potential effects, and then monitor the results of actual operations.

To deal with the increased complexity, time and scope of such planning methods, Dr. Allen introduces new conceptual models or paradigms in this text to help campaign planners comprehend the space of options and issues in the planning process. He provides a comprehensive overview of the planning process that integrates traditional planning methods and advanced concepts into a cohesive introduction to the subject.

More than just a philosophical discussion, this text defines the first principles of planning, describes techniques to explicitly represent and simulate plans, and illustrates the methods with screenshots of actual planning tools. The book describes how these new approaches have been implemented in recently fielded planning tools, and how the methodologies described in the book will continue to evolve over time.

We are fortunate to learn from a pioneer and leader in this exciting field. Beyond sharing his lessons learned from nearly a decade of research and development, he describes his perspective of future challenges to stimulate the next generation of R&D.

Ed Waltz[1]
Ann Arbor, Michigan
October 2006

---

1. Ed Waltz is the editor for the Artech House series of texts on Information Warfare, and lectures internationally on the subjects of intelligence and information operations. He is the chief scientist of the Intelligence Innovation Division of BAE Systems, Advanced Information Technologies.

Part I
Information Operations Planning Context

1

Information Operations and Its Role in Security and Conflict

How do you know what you know? What is the truth, and how is it determined?

—Dr. Simon Ball

1.1 The Purposes and Principles of Information Operations: What Do You Think You Know?

Outcomes are based on actions; actions are based on decisions; decisions are based on perceptions; and perceptions are based on information. But how do you know that your perceptions are true? How do you know if the information is accurate and up-to-date? How do you know whether your decisions are available to those who are to implement them and not available to those trying to stop them? How confident are you in the perceptions upon which you base your decisions? These are the types of questions associated with *offensive* and *defensive* information operations (IO).[1]

Although the leading quote is from a course on epistemology, these questions are very relevant to the practical application of military information operations and the planning processes and tools that support them. How do you know that what you think you know is actually true? How do we cause the other side to "know" what is not so? How do you get the other side to doubt what they

---

1. We will discuss the issues associated with the third area of IO, *influence*, later.

think they know, even if it is true? How can we be sure that what we think the enemy knows is true, and that they are not just fooling us instead? How can you be sure that the enemy does not know what you know?

If our side can answer all of theses questions favorably, while our opponents can only answer these questions unfavorably, then our side is said to have *information superiority*. Information superiority leads to decision superiority, which leads to dominance across the spectrum of operations. *Information inferiority* leads to traps or doubt, and doubt leads to decision delays and undermines commitment to decisions when initial results are not as expected.

Like any endeavor, military operations require the availability and integrity of information to support good decisions and plans. The confidentiality of information is also important. You do not want the other side to know what you are attempting to do; otherwise, they can take steps to negate your desired outcomes and/or the actions leading up to these outcomes. The confidentiality, integrity, and availability (CIA) of information are three of the cornerstones of information assurance, but there is a fourth factor to consider in information operations as well. That fourth factor is confidence.

If you are confident that your perceptions are accurate, then you are likely to be confident in your decisions, confident in your actions, and confident that the desired outcomes will be achieved despite early adverse outcomes. However, if one has significant doubts about one's perceptions, that doubt leads to decision delays and also undermines commitment to seeing decisions through when initial results are not as expected.

In addition to offensive and defensive IO, there is a third and equally important area called *influence* operations. For influence operations, IO planners are interested in a different but related set of questions. How do you convince a group to accept your message over an opposing group's message? What information, approach, and delivery mechanism will lead to our message gaining the attention and belief of a specific group? How do you know when that group has actually accepted your message? For the purpose of organizing this book, we will use the categorization of offensive, defensive, and influence to define three broad areas and concerns of IO planning.[2]

Compared to other military endeavors, information operations as its own field is relatively new. Even though deception and intelligence have been used throughout the history of warfare, the importance of IO across all aspects of

---

2. Categorizing IO into attack, defense, and influence in this book makes it easier to explain the issues and concepts associated with planning each type of operation. If the book followed the organization of the five core categories of IO currently defined by DoD (CNO, EW, PSYOP, MILDEC, and OPSEC), as well as the related and supporting categories, there would be substantial duplication of material in many chapters. See Chapter 4, Figure 4.1, for the mapping of DoD's categories to this book's categories and chapters.

military and political operations has reached new heights in the Information Age. (This is discussed further in Chapter 2.) Because of its newness, the military science for IO is not as well developed as the sciences of warfare in air, land, and sea. The U.S. Department of Defense (DoD) defines IO in Joint Publication 3-13 (JP 3-13) as follows [1]:

> The integrated employment of electronic warfare (EW), computer network operations (CNO), psychological operations (PSYOP), military deception (MILDEC), and operations security (OPSEC), in concert with specified supporting and related capabilities, to influence, disrupt, corrupt, or usurp adversarial human and automated decision making while protecting our own. [1]

The 1998 version of JP 3-13 defined information warfare (IW) as "information operations conducted during times of crisis or conflict to achieve or promote specific objectives over a specific adversary or adversaries." [1] While IO is applied during both peacetime and conflict, IW tends to be applied during conflict or crisis. This is because the techniques of IW are assumed to be the employment of coercive force and will usually be perceived as a hostile act. Since the border line between IO and IW was not well defined, IW was deleted from the 2006 version of JP 3-13 [1].

While information operations is a field that is full of depth and complexity (as described further in Chapter 3), there are useful ways of thinking about and categorizing IO that can help planners and warfighters wrap their arms around this complex space and consider a relatively small set of options at each decision point. The purpose of this book is to help provide an accurate picture of what IO is about and describe recently developed techniques to make the space of IO options manageable to the planner and warfighter. Since much of the success of command and control lies in creating a shared vision and perception [2], this book is intended to provide a starting point for creating that shared vision and perception of IO, and implementing that vision in modern IO planning tools.

Just as important is clearly defining what IO is not, which we present later in this chapter. Before that, however, we present a set of examples of IO used in recent conflicts that have been gathered from unclassified sources. This serves to introduce the reader to some of the types of effects information operations have actually been able to achieve.

1.2 Examples of IO Implementations in Recent Conflicts

Information operations are being undertaken daily by many groups around the world, especially in the current war on terrorism. The following set of examples, presented in chronological order, helps introduce the reader to some concrete examples of IO and how it has been used in the last decade or two. These real-world examples are limited to what can be gleaned from open sources, and do not begin to address the full scope and detail of real-world IO applications or the methodologies and tools that may have been applied in the last two decades.

1.2.1 Operation Desert Storm (ODS)

ODS has been called the first information war, but the number of examples that can be referenced is fairly small. For an example of what DoD would call an information operation, CENTCOM published on the tenth anniversary of Operation Desert Storm a set of interviews from key participants and other sources. The article on IW stated [3]: "Another aspect of Desert Storm IW involved the use of electronic tags placed earlier in U.S. exported computers. These tags, once activated, helped reveal computer locations within Iraq and their probable military applications. The computer tags, employed originally for export control and intelligence applications, provided yet another facet of IW."

1.2.2 Serb IO in the Kosovo Campaign

A number of articles have been written about how the Serbs, although at a serious military disadvantage compared to the alliance, performed very well in the information war, as can be seen in the following three examples.

Military Deception

The Serbs were experts at creating target decoys. They knew the limitations of Allied overhead surveillance times and placed actual targets where they could be seen. Then they swapped them with dummy targets or unserviceable vehicles at nightfall, since the Allies restricted themselves to only certain windows to bomb targets [4]. Besides numerous physical decoys, the Serbian military made extensive use of false message traffic to cause the Allies to bomb empty or nonexistent targets. The Serbs learned from the Iraqis' experience in Operation Desert Storm and understood that their radio message traffic was subject to fairly extensive eavesdropping. Therefore, the Serbs generated a number of false radio messages specifically designed to cause the Allied aircraft to bomb empty targets, or targets that would cause collateral damage. The Serbian use of false radio traffic was so extensive that it was difficult to tell whether any Allied radio intercepts led to any strikes that caused any damage to enemy forces [5].

Propaganda

In terms of the public opinion competition, the Serbs kept the Alliance on the defensive, retaining the initiative by accusations of war crimes and atrocities throughout the conflict. Regardless of the facts on the ground, the Serbs exploited international media to bombard the Allies with a constant stream of accusations related to any collateral damage that might have been caused during bombing [4]. Moreover, the Serbs and the Russians together after the conflict brought claims against the United States and NATO for war crimes based on the claim that bombing from such a high altitude led to wanton damage and loss of civilian life. Although disproved, the accusation continued the momentum of Serb accusations against the Alliance even after the Kosovo conflict was over [personal communication with Jeff Walker, October 7, 2003].

Open-Source Intelligence

The Serbs had agents or sympathizers with cell phones sitting on lawn chairs at the end of the runways at Aviano and other Allied airbases. When aircraft would take off, the number and type of aircraft launched were called in to Serb intelligence. By use of simple flight time calculations based on the types and number of aircraft, the Serbs were able to determine with fairly good accuracy when Allied formations would be over various target areas. This open-source early warning system probably contributed greatly to the lack of serious damage against the Serbian military caused by the air raids [6].

In addition, it was known that the Russians sent electronic eavesdropping ships to the area, and they may have shared intelligence with the Serbs [7]. Since the Allies did not have good secure communications interoperability, this "forced reliance on non-secure methods that compromised operational security" [8]. Communicating military operations in the clear is almost the equivalent of open source intelligence. If Russia or other nations were, in fact, listening in to unencrypted NATO communications, it was the equivalent of Serb agents directly listening in to those same communications, albeit with a slight delay time.

1.2.3 Allied IO in the Kosovo Campaign

Conversely, the Allies also appear to have successfully applied IO as leverage against the Serbian decision makers, although the effectiveness against the Serbian population remains undetermined.

PSYOP Against the Serbian Leadership

In military strategy formulation, it is important to determine the centers of gravity of the enemy. In the case of Serbia, this was initially considered to be the Serbian military. However, as the campaign progressed, it became very clear that the real center of gravity was the megalomania and popularity of the top leader of Serbia, Slobodan Milosevic. Finding leverage points over Milosevic was initially difficult. However, the Land Information Warfare Agency (LIWA) helped identify the fact that the Milosevic family and the families of his cronies had their wealth contained in a relatively small set of industries in Serbia. Once these leverage points were identified, it was possible to place significant pressure on Milosevic and his crony structure [9]. That pressure and the lack of Russian support on the international arena were the leading causes of Milosevic's decision to back off from Kosovo [8].

The IO methodology used in this case involved thinking through what was important or valuable to the leadership of the opposition. Since the use of Allied ground forces had been precluded, it soon became clear that the only thing that could cause Serbian withdrawal from Kosovo was a decision from Milosevic [7]. Therefore, pressuring Milosevic to make that decision revolved around affecting what mattered to him—his wealth and the wealth of his family and friends.

PSYOP for the Serbian People

The Allies, and especially the United States, made a number of public statements that the conflict was against Milosevic and not against the Serbian people. This was intended to help defer the solidification of public support in Serbia behind Milosevic and to give the Serbians a way to separate themselves from the perpetrator. The main broadcast to the Serbian population on television was timed to occur during their most popular TV show. This turned out to be both good news and bad news. The good news was that the message reached the largest group of population at one time. The bad news was that it "spoiled" their most popular TV show for the week, leading to substantial resentment [personal communication with Jeff Walker, October 7, 2003].

1.2.4 Internet Conflicts

Due to its interconnectedness and interdependencies, the Internet has also been used as part of the battleground in national conflicts. Pro-Serbian hackers attacked U.S. and NATO Web sites, pro-Palestinians and pro-Israelis have attacked each others' sites, while hackers supporting India versus Pakistan, China versus Taiwan, and China versus the United States have done the same [10]. Note that in each case, the attacking nations were not using national hacking resources. Instead, hackers from around the globe became involved in a form of "hactivism" [11]. At one point, the Palestinian-Israeli cyber conflict of 2000–2001 had participants from Germany, Brazil, the United States, Saudi Arabia, Pakistan, and Lebanon, as well as from Palestine and Israel. The U.S.-China cyber skirmish over the EP3 incident in the early 2000s attracted hackers from Japan, Indonesia, Korea, Malaysia, Argentina, Brazil, India, Pakistan, and Saudi Arabia, as well as the United States and China [12].

In most cases these attacks were nuisance "raids" equivalent to painting graffiti on Web sites [13]. However, in some cases, the degree of damage was more significant, including the deletion of data, days of lost sales, and even a claim by one group that the 8% drop in the Israeli stock market was caused by fears generated by these cyber attacks [14]. Probably the most chilling event was when Chinese hackers gained access to a California electric power company's test and demonstration network, even though the penetrated site was not connected to the actual power grid. The site's defenders assumed that by not listing the test site's URL address in the domain name server, it could not be found and was therefore secure [15]. The Chinese hackers used a brute force search of IP addresses based on a "starting block" of addresses purchased by the company. It did not take long for the Chinese hackers to find a "live" IP address not listed in the domain name service, whereupon they immediately attacked and penetrated the otherwise undefended site [16].

Web site security software is now available that helps Web sites better protect themselves from direct defacement, but technological advances continue to enable hackers to find new ways to exploit Internet vulnerabilities [17]. For example, the FBI broke up a planned New Year's Day denial of service attack against the whole Internet planned for January 2001 [18]. More than 500 servers had been infected by the hackers to act as zombies, scheduled to simultaneously flood the Internet with enough traffic to grind it to a halt [19]. Large-scale denial of service attacks via a large number of compromised servers are still a valid threat against the Internet and therefore against international commerce.

1.2.5 Allied IO in Operation Iraqi Freedom

Just over a decade after Operation Desert Storm, the Allies were again fighting against Saddam's Iraq. This time, the Allied IO capabilities were much more extensive than in the previous conflict.

Countering Global Positioning System (GPS) Jammers

An important analysis of the Iraqi GPS jamming capabilities was performed prior to the start of hostilities. The U.S. Air Force estimated likely locations for the Iraqi GPS jammers, and used the GPS Interference and Navigation Tool (GIANT) to determine the effects the jammers would have on the release and trajectory of GPS-guided munitions. The analysis showed that the effect would be minimal given the scenario's delivery conditions. After the major combat operations were concluded, a postmortem was performed on Iraqi jamming capabilities (actual location and use) and the observed effects on delivered munitions. The predictions of GPS jammer locations matched well with the actual locations, and the observed effects on the munitions appeared to match the predicted effects (none or minimal) [20].

Transmitting on Enemy Network Frequencies

The Coalition went beyond just jamming to actually using enemy communications networks to transmit U.S. PSYOP messages. "The EC-130H Compass Call communications jammer, for example, for the first time not only blocked enemy communications but was also able to take over Iraqi tactical radio networks to broadcast U.S. psychological operations (psyops) messages" [21]. The Navy also transmitted PSYOP messages from aircraft and ships. "The task was performed with 'no significant modifications' by using an interface developed by the Fleet Information Warfare Center" [21].

Communications Herding

According to [21], "The U.S. conducted 'communications herding,' whereby most frequencies were jammed, forcing Iraqi military to broadcast from a small set of other frequencies that were more easily disrupted or exploited for intelligence. One such instance was the destruction of Iraq's fiber-optic communications before the war, which channeled Iraqi forces into using more easily intercepted high-frequency radios."

Computer Network Attack

According to [21], "Classified computer network operation tools were used in OIF, primarily to send messages directly to key commanders as well as help judge their effectiveness of the larger psyops or perception management campaign on specific targets …. However, Maj. General Gene Renuart, Central Command's (CENTCOM's) Director of Operations told JDW that computer network operations 'were only a small part of our (IO) operations,' mainly because there were only about 15,000 computers with Internet access in the country at the time and because that access was closely monitored by the Ba'ath regime."

1.2.6 IO After the Liberation of Iraq

Winning the peace and attaining stability rely heavily on information operations and intelligence operations for success. Just finding the enemy is difficult, while figuring out how to encourage hostile ethnic groups to work together is even more difficult.

Communications Monitoring

When the Hussein sons were cornered in the house of their last shoot-out, they made a number of panicked calls to supporters. Since the house was under electronic surveillance, this was an intelligence windfall for U.S. forces [22]. Everyone who the Hussein brothers called for help became immediate suspects as possibly supporting the insurgency.

Exploiting Western Media

The forces opposed to the U.S. liberation and reconstruction of Iraq are very good at using Western media to focus attention on their effort. Saddam Hussein had, until captured, delivered a regular set of audiotapes to news media, especially Al Jazeera, to remind his followers of his continued presence and to sustain the effort. (Al Zarqawi and Bin Laden have used the same technique.) Getting and sustaining attention are two of the most important aspects of political struggles. The Western media's tendency to focus on death and destruction as the top news items plays very well into their efforts.

Western media naturally tends to follow the following criteria to define the top news stories of the day:

1. Deaths, and the more deaths, the more coverage;
2. Casualties, and the greater number, the greater coverage;
3. Damage, and the more damage, the greater the coverage;
4. Unrest, and the larger the unrest, the greater the coverage;
5. Speeches, even by the U.S. president.

For example, most of the reports on October 27, 2003, about four Baghdad car bombs focused on the number of people killed [23]. The fact that most of the car bombs did not reach their intended targets, and that the "bombed" buildings were damaged but not destroyed because the cars were stopped by barriers and/or guards, received little attention. The Western media did not focus on the failure of the terrorists to reach their intended targets, nor the fact that the terrorists wantonly killed numerous innocent Iraqis whom they are claiming to be fighting to protect. The emphasis was entirely on the fact that the explosions occurred and people were killed and injured, which is great press for the terrorists.

The *Washington Post* interviewed 100 veterans returning from Iraq, including their views of the war versus the media coverage of the war. The summary article stated [24]:

> But it was not bad in the ways they see covered in the media—the majority also agreed on this. What they experienced was more complex than the war they saw on television and in print. It was dangerous and confused, yes, but most of the vets also recalled enemies routed, buildings built and children befriended, against long odds in a poor and demoralized country. "We feel like we're doing something, and then we look at the news and you feel like you're getting bashed." "It seems to me the media had a predetermined script." The vibe of the coverage is just "so, so, so negative."

As a result, Western media helps fuel the political strength of the opposition because the opposition understands what drives the Western media's need for attention from their consumers. Also, just as naturally, the progress of reconstruction in Iraq, which has a much longer time frame in terms of success and ultimate effect, is given little attention by the media. This isn't so much a liberal media bias against the Administration, but a natural tendency for automatic prioritization of attention-getting news stories within Western media. (This will be discussed more thoroughly in Chapter 6.)

Foiled Propaganda

Conversely, not all opposition propaganda attempts are successful. As reported in *Time*, an opposition member posing as an informant reported men and weapons at a nearby mosque being prepared for an attack. The U.S. forces prepared to raid the mosque, but first attempted to corroborate the information. Shortly before the raid was planned to occur, it was cancelled for lack of corroboration, and the realization that this was likely a propaganda effort by the opposition to film a U.S. raid on a peaceful mosque at prayer during Ramadan [25]. The method of thought (seeking corroboration) and the awareness that the opposition is trying to set up such incidents helped prevent an international media coup for the opposition.

Trying to Incite a Civil War

On February 22, 2006, terrorists destroyed the Askariya shrine in Samarra in an effort to inflame ethnic and religious tensions in Iraq into a full-scale civil war [26]. While this brutal action did substantially increase the violence between Sunni and Shiite groups in particular, it did not generate the war the terrorists wished to initiate [27]. What is amazing is that the terrorists received little backlash from this heinous crime, either within Iraq or internationally. Instead, some in Iraq and Iran blamed the United States and "Zionists" [26].

1.2.7 Cultural Messages via the Media

The Information Age has provided the opportunity for multiple cultures to present their views at the same time to wide audiences. Sometimes these messages are understood only by one part of the audience, which allows for messages to be sent to one group while the other group remains ignorant. Two recent examples underscore the power of sending a culturally based message "in the open."

The first example occurred on October 15, 2004, when Dr. Condoleezza Rice presented the U.S. case to the Muslim world on Al Jazeera [28]. At the start of the event, the Al Jazeera interviewer put out his left hand to Dr. Rice, who took it with her right and smiled at the gesture. What was apparent to any Arab and many Muslims is that the interviewer had just given Dr. Rice a severe insult to which she appeared unaware. Offering to shake hands with the left hand is a strong insult in the Arab world [29]. Whether Dr. Rice was ignorant of the insult or simply chose to ignore it is irrelevant. While she may have been instructed to ignore such cultural insults, the message was clear to all Arab observers, and a coup for Al Jazeera.

The second example is a little more complex. In 2005, when Interim Iraqi Prime Minister Iyad Allawi came to visit the White House and to address Congress, a joint press conference was held in the Rose Garden. At one point, President Bush held out his right hand to Prime Minister Allawi at the adjacent podium. Allawi twisted himself around to ensure that he used his left hand to shake President Bush's right hand [30]. As it turns out, Allawi had broken his right hand a month previously, reportedly by slamming it down on a desk during the fighting in Najaf [31]. So there was a valid reason for President Allawi to shake with his left hand. But what was the message sent to the Arab world? Was this the message he wanted to send? Was it the message we wanted him to send? Whatever the purpose, the message was clear to all Iraqis: When Allawi used his left hand to shake President Bush's right hand, it probably played well in Iraq. Intentionally or not, he showed his countrymen that he distanced himself from Bush, while not obviously shunning the U.S. president in front of his own people.

Clausewitz stated that war is the continuation of politics by other means [32]. As many of these examples have demonstrated, in the Information Age, the connection between war and politics is more closely coupled than ever before. A tactical event involving one or a few soldiers can have international repercussions in the political realm. Note the Abu Ghraib abuse scandal, the shooting of a wounded insurgent in a mosque, or the draping of the American flag over a statue of Saddam Hussein. Conversely, political statements can significantly affect military operations. For example, President Clinton's statement at the start of the Kosovo conflict that no troops would be placed on the ground provided significant flexibility to the Serb forces and greatly hampered Allied military options [7].

Moreover, because of the interconnected and multipolar nature of the modern world, conflict in the Information Age is no longer just two-sided but multisided. Fighting for democracy appeals to Americans and many Western nations, but is somewhat threatening to U.S.-friendly governments that are monarchies or oligarchies [33]. Messages designed for one audience have to be carefully crafted to avoid generating negative reactions in other audiences. While much of the attack and defense aspects of IO continue to focus primarily on two-sided conflict, *the realm of influence operations tends to focus substantial effort on the multisided nature of the modern world.*

1.3 What IO Is Not

One useful way of describing something is to also clearly define what it is not. Since all learning is a process of comparing and contrasting something new to something familiar, we will spend some time describing what IO is not.

A number of definitions and examples of IO have been used over the years, and most of the ones listed below suffer from the "blind men describing an elephant" syndrome. One grabs the tail and says it is like a cord; one grabs a leg and says it is like a tree; one touches the side and says it is like a wall; one grabs an ear and says it is like a sheet; one grabs a tusk and says it is like a spear; and one grabs the trunk and says it is like a snake. Although each one has described a local feature of the whole beast, none have described the elephant as a whole, or the "holistic elephant."

In a similar manner, most of the misconceptions and poor examples of IO have come from focusing on a small subset of IO, resulting in a definition that is ultimately false, or an example that is misleading at best. We refute some of the most common incorrect definitions by clearly stating what IO is not in each of the following examples.

IO Is Not Just Slowing Down an Enemy's "OODA Loop"

Colonel Boyd described the "observe, orient, decide, act" (OODA) loop originally for air-to-air combat, but it appears to be applicable to a wide range of situations, especially competing decision processes in a combat situation [34]. For example, in France in 1940, the German blitzkrieg consistently performed all steps of their OODA loop significantly faster than the French in every case. Because the French were always slower to react to any threat, this led to paralysis of the French forces [35].

Due to the popularity of the OODA loop concept, some have attempted to define IO as simply the slowing down of the enemy's OODA loop in order to obtain and sustain an OODA loop advantage in combat [36]. While slowing the enemy's OODA loop is one way to use IO, there are other ways to use IO that don't delay the enemy's OODA loop, or that make the enemy's OODA loop irrelevant to the friendly objective.

For example, if the friendly side has successfully convinced the enemy that a friendly deception plan is the real plan, then the friendly side does not want to delay the enemy walking into that trap. As Napoleon stated [37], "Never interrupt your enemy when he is making a mistake." Since the enemy's observation is false and his orientation is fitting into your plan, you want the enemy to decide and act as quickly as possible on that false information and walk into your trap. In this case, slowing the enemy's OODA loop is counter to the desired outcome. The longer the delay, the more likely the enemy will change course and avoid walking into the trap regardless of whether it was even detected.

IO Is Not Just Influence Operations

Another common misperception is that IO encompasses only influence operations that affect people's minds. The phrase "IO is the name, influence is the game" is false (by being too limiting), but has appeared frequently in the psychological operations (PSYOP) community [38]. It is true that a major portion of IO involves influence operations that target select groups for the delivery of messages to change beliefs and behaviors. But influence operations ignore the technical aspects of IO that act against opposing information and information systems and help protect friendly information and information systems.

An interesting side effect of this misperception appeared in portions of the military testing community. Some of the developmental and operational testers believed that since IO only applied to influence operations, it did not apply to the more quantitatively focused testing community. In reality, however, the need to test new systems and procedures in an IO environment is more necessary than ever. The U.S. military's increasing reliance on commmercial communications and computation technologies increases the vulnerability of our forces simply due to that increasing reliance. The testing community needs to better address both the offensive and defensive sides of IO in their testing environments, and some preliminary steps have been taken in that direction.

IO Is Not Just Special Technical Operations (STO)

The converse to the previous misunderstanding is applicable to the military community associated with technical operations. The community that is focused on the technical aspects of information storage, flows, and processing tends to forget that the ultimate aim of affecting information is to affect enemy decisions [5]. Outcomes rely on actions, actions rely on decisions, decisions rely on perceptions, and perceptions rely on the available information. If that information is incomplete or distorted by the opposing side (more so than caused by the normal fog of war), then the decisions will be based on false or distorted perceptions, leading to the probability of poor decisions. Focusing on the technical aspects to the exclusion of the cognitive aspects will miss the essential interaction among the technical and cognitive synergy that can be used to discourage, deceive, or deter the enemy.

This leads to another aspect of IO—you can't *guarantee* that the enemy will decide and act as you desire. Even if you have the perfect deception plan and have spoofed all of their information systems, the enemy may *still* make a decision that is contrary to where you have been trying to lead him. The best one can do is to increase the probability that the enemy will make the desired decision as guided by your offensive IO successes [39]. This is described further in Chapter 6.

IO Is Not Just Electronic Warfare (EW)

A subset of the previous misperception is that IO is just EW by another name. This definition misses most of the influence and computer network operations aspects of IO. The specific claim of ownership is that IO is encompassed by EW because everything that involves the electromagnetic (EM) spectrum belongs to EW. Traditionally, EW referred to the radiated electromagnetic spectrum.

This effort to claim the complete EM spectrum for EW is flawed on several counts. First, according to JP 3-13, *Information Operations*, the five core IO capabilities are electronic warfare, operations security (OPSEC), psychological operations (PSYOP), military deception, and computer network operations (CNO). Second, according to Air Force Doctrine, the three operational functions of IO are electronic combat operations (where traditional EW lies), network combat operations, and influence operations.

Therefore, the thrust of the EW claim to the full EM spectrum appears to be an effort to control CNO and possibly the OPSEC core capabilities as well. However, there is much more to computer network operations than simply operating in the EM spectrum. First, for example, social engineering—nonelectronic ways to gain access to computer networks—is completely separate from the EM spectrum. Second, focusing on the EM spectrum misses the longer time frames involved in CNO and IO. For example, placing a Trojan Horse virus for later access, or setting up for time-delayed launching of software or physical actions, does not benefit from focusing on just the EM spectrum. Third, physical access to, or interference with, a computer network is part of the CNO charter, yet that also lies beyond the EM spectrum. Fourth, although parts of military deception can be performed in the EM spectrum, many other parts cannot. Lastly, only a very small portion of PSYOP and other influence operations involve the EM spectrum.

To allow EW to claim that it owns the EM spectrum, and therefore any IO activities associated with the EM spectrum, misses most of the essential synergy between actions taken in and outside the EM spectrum to achieve IO mission objectives. Moreover, focusing on the EM spectrum for guiding IO is like focusing on sea state as the basis for all naval operations. Sea state can influence cooperative engagement, but cooperative engagement goes far beyond the scope of the definition of sea state. In a similar manner, humans conduct electrical currents between neurons, and to claim that all human-related activities are based on the ownership of the EM spectrum is ludicrous.

The reader needs to be aware that the definition of IO is not just an academic exercise, but becomes the basis for organizational roles, authority, and budgets. As a result, we need to be sensitive to definitions that are primarily aimed at claiming ownership. Fortunately, both Joint and Air Force doctrine agree that EW is a core element of IO and that IO is not encompassed by EW.

IO Is Not Just Information Assurance

Information assurance (IA) is not completely encompassed by IO, nor is IO encompassed by IA. The existing overlap among definitions of IO and IA are recognized by DoD's new IO definition (Chapter 4). IO, by definition, involves an adversarial situation, where humans or manmade systems are designed to attack and defend, or compete against each other in the realm of influence. IA, however, is designed to ensure the confidentiality, integrity, and availability (CIA) of information regardless of the source of the threat to that information. IA attempts to provide that CIA of information whether it be threatened by an adversary, nature, technical failure, or human stupidity. For example, a backup server used as a "hot spare" is one way to ensure continued operations, regardless of the cause of the primary server going down. Since no enemy action is being thwarted in most commercial settings, information assurance includes all of the preventative and remediation steps necessary to sustain uninterrupted operations even with no enemy threat.

Although all of the steps that can be taken in IA can be useful in protecting against, or mitigating against, the effects of an enemy attack, they can also be used across a wider range of activities, especially nonmilitary ones. As a result, the DoD definition of IO does not completely encompass the definition of IA, nor should it. IA is something that companies and agencies should strive for to ensure continuity, integrity, and availability to their necessary information, regardless of whether there is a human or cyber threat. Conversely, IA does not encompass the attack or influence aspects of IO.

Cultural Misunderstandings of IO

Overall, a lot of the misunderstandings about IO occur simply because folks are used to one aspect of IO, especially if they came from a community that had its own cultural approach, such as operational art, PSYOP, STO, or EW. (These are not the only communities that may have a limited view of IO and only serve as an example.) When one is familiar with PSYOP, IO looks like PSYOP. When one is familiar with EW, IO looks like EW. Yet IO is more than what can be encompassed by just one of these communities.

For example, the original version of Air Force IO doctrine consciously chose to map IO elements as parallels to offensive and defensive counterair in order to help get the Air Force used to the idea of using something with which they were familiar. This follows the principle of comparing IO to something familiar to help the audience understand the topic. The problem is that without a broader understanding of what else IO is like and what it is not like, the initial "familiar" definition becomes a stumbling block to fully understanding what IO is and is not. The contrasting part of comparing and contrasting is still needed to complete the understanding. As a result, the Air Force Doctrine for IO still does not mesh well with the Joint definition of IO, although the Air Force definitions are getting closer to the Joint definitions over each iteration of AFDD 2-5. The Air Force still teaches and follows its own doctrine for IO, even though all parties have agreed that the Joint doctrine is dominant when in conflict with Air Force doctrine.

In the opinion of this author, there are two primary reasons why many of the misunderstandings of IO have perpetuated for so long. First, people focus on what is familiar in the community from which they came, because they understand it and are comfortable with it. Their worldviews tend to be limited to that community, which naturally lacks the exposure necessary to expand to the other communities that are more familiar with the other aspects of the IO elephant. Second, limited definitions of IO are often used to enhance one organization's fiscal and bureaucratic turf wars. It is in these organizations' self-interest to focus on only a portion of the definition. As a result, the various parties involved may be slow to reach or accept an agreed-upon definition of IO simply because each is concerned that a common definition may be more favorable to one Pentagon office than another.

There is hope, however. As more personnel from more offices have begun to recognize the value and scope of IO, and as some of the turf battles among offices attempting to dominate IO have begun to sort themselves out, we may be moving closer to a commonly accepted definition of IO. The new DoD definition of IO in the DoDD3600 series is a step in the right direction, as described in Chapter 4. Before that, we will discuss in Chapter 2 why information operations are more important now than ever before due to the emergence of the Information Age and our dependence on its technologies. Meanwhile, the next section will discuss how IO relates to the emerging and evolving method of thought called *effects-based operations* (EBO).

1.4 The Role of IO in Effects-Based Operations (EBO)

IO has a role to play in conflicts, and in security operations in general. Information operations do not occur in a vacuum, nor is it the only tool available to a nation. IO is only one part of a nation's toolset of methods to influence or coerce other nations. The larger, all-encompassing definition of the space of national capabilities developed over the last decade or so is called effects-based operations (EBO). The U.S. Joint Forces Command, U.S. Strategic Command, and some of the U.S. Military Services have been experimenting with and defining EBO. The U.S. Air Force definition [40] of EBO is "actions taken against enemy systems designed to achieve specific effects that contribute directly to desired military and political outcomes."

In practice, EBO is the employment of all instruments of national power (diplomatic, informational, military, and economic—or DIME) against opposing political, military, economic, social, information, and infrastructure (PMESII) capabilities to create a desired effect or end state. Since EBO tends to be an all-encompassing method of thought, then IO (along with all military, economic, and diplomatic capabilities) should be employed as part of the overall EBO-guided campaign. Since information is one of the DIME elements, and information is one of the PMESII targets, information operations clearly lies within effects-based operations.

IO planning is also not performed in isolation. The objectives, tasks, and actions of IO are (or should be) designed to follow some purpose or mission and contribute in some way to the achievement of that purpose or mission. In the United States, it is the civil authorities that define political objectives and the role the military plays in achieving them. In the modern world, where winning the war is often easier than winning the peace, the role of nonmilitary capabilities, whether wielded by the military or some other agency, is more frequently being considered as the preferred tool of choice.

EBO and IO both attempt to expand the repertoire of traditional military capabilities beyond just blowing things up (which is sometimes called a "kinetic solution"). Nonkinetic solutions include the ability to disrupt the enemy's electrical power, force a reaction by denying an economic resource, or one of many other ways to get the enemy to respond the way you want without killing him. Note that EBO includes the kinetic solutions, but also includes a wide range of nonkinetic solutions. In addition, effects-based planning can be used to support stability operations and nation-building, where the intent is to do good things to the nation and not just disruptive things that are planned during conflict.

Although EBO expands the range of capabilities available to the political and military decision makers, it also greatly increases the political and military demand for a wide range of information not usually handled by traditional military, or even national, intelligence. For example, in order to know the best place to disrupt a network (e.g., transportation, communications, fuel, or electrical), a great deal must be known about the target network and its backup systems and recovery procedures before the desired effect can be obtained. This level of information requirements is far beyond traditional military intelligence requirements. As a result, the demand for information about a large number of physical and nonphysical items, as well as their known or likely interactions, is required. The result is a *huge* demand for information. (See Chapter 10 on the logistics of IO.) The initial effort to provide this increased level of information for EBO was called Operational Net Assessment (ONA) by the USJFCOM Joint Futures Lab, and was designed to be the data source for EBO planning. "ONA is an integrated plans, operations, and intelligence process" [41]. Since IO planning is an element of EBO planning, IO was supposed to tap into the ONA pool, define its own information requirements, and help prepare the space for future operations. However, during USJFCOM's transformation experiments, the initial ONA approach was found to be unwieldy and not sufficiently adaptive. As a result, USJFCOM is looking for an alternative approach to preparing all of the required information to support EBO in general and IO in particular.

Regardless of which method is eventually adopted, the United States cannot afford to wait until a crisis begins to collect all of this required information. The necessary data must be collected in peacetime on a nearly continual basis. Without *information readiness*—having the information you need before you need it—our forces will not be ready to undertake EBO or IO. (See Chapter 10 for more on information readiness.)

IO and EBO planners require more than just information about current status. They also require estimates of likely future status across a range of future planning horizons. Predictive battlespace awareness (PBA) is designed to use a wide range of tools to take current information and predict future status—especially enemy intent and future status. (Although PBA is the term the U.S. Air Force uses to provide predicting intelligence and assessments, each of the Services has its own version of it.) Since IO benefits from being able to predict the effects of its actions, likely enemy responses, and future enemy activities, PBA supports EBO and IO as well [42].

Overall, information operations comprise one of the four major instruments of national power (the DIME described earlier), as shown in Figure 1.1. (Note that while the U.S. military currently performs many of the nation's information operations, IO is considered under the information instrument of power vice the military instrument of power.) IO can be used against the enemy's information resources, or against any of the other five PMESII aspects of a nation, including directly influencing the population. IO can be used by itself or in synergy with the other three instruments of national power, both in a supporting role and in a "supported by" role. Moreover, IO potentially provides

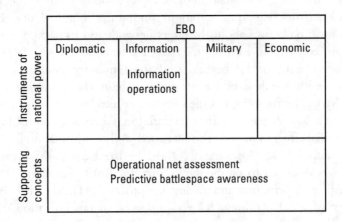

Figure 1.1 IO is part of EBO and supported by ONA and PBA.

a large repertoire of nonlethal or less lethal options to a nation attempting to influence or coerce a potential or actual opponent.

The promise of IO is great, while its accomplishments to date have been mixed. Part of the reason for the mixed results is that the full capabilities, constraints, and requirements of IO are not well understood. Another reason is that the relative newness of IO compared to the other instruments of national power has naturally led to growing pains in terms of concepts of operation, doctrine, materiel acquisition, training, planning, and execution. This book is designed to help present what might actually be possible in the realm of implementing information operations in the next decade or two, and to help the reader wrap his or her arms around how one can plan, and monitor the execution of these plans, in such a broad space of potential capabilities.

References

[1] Joint Publication 3-13, *Information Operations*, February 13, 2006.

[2] Kahan, James, D. Robert Worley, and Cathleen Statz, *Understanding Commanders' Information Needs*, RAND Report R-3761-A, 1989.

[3] CENTCOM, *Desert Shield Desert Storm, the 10th Anniversary of the Gulf War*, U.S. Central Command, published by Faircount LLC, 2001, http://www.faircount.com.

[4] Thomas, Timothy, L., "Kosovo and the Current Myth of Information Superiority," *Parameters*, Spring 2000, http://www.carlisle.army.mil/usawc/parameters/00spring/contents.htm.

[5] Keys, Maj. Gen. Ron, *Future War and the Lessons of Kosovo*, Briefing, January 10, 2000.

[6] Rosenberg, Robert A., "Incidental Paper from the Seminar on Intelligence, Command, and Control, Center for Information Policy Research," Harvard University, August 2002, http://www.pirp.harvard.edu/pubs_pdf/ rosenbe%5Crosenbe-i02-1.pdf.

[7] Levien, Frederic H., "Kosovo: An IW Report Card," *Journal of Electronic Defense*, Vol. 22, No. 8, August 1, 1999.

[8] Cohen, Sec. Def. William S., and General Henry H. Shelton, "Joint Statement on the Kosovo After Action Review," *Defense Link News*, October 14, 1999, http://www.defenselink.mil/news/Oct1999/b10141999_bt478-99.html.

[9] Arkin, William, and Robert Windrem, "The Other Kosovo War," August 29, 2001; http://www.msnbc.com/news/607032.asp.

[10] Hershman, Tania, "Israel's Seminar on Cyberwar," Wired.com, January 10, 2001; http://www.wired.com/news/politics/0,1283,41048,00.html.

[11] Kirby, Carrie, "Hacking with a Conscience Is a New Trend," originally appeared in *San Francisco Chronicle*, November 20, 2000; reported on Infowar.com, 11/24/2000; http://www.infowar.com/hacker/00/hack_112400a_j.shtml.

[12] Allen, Patrick, and Chris Demchak, "The Palestinian/Israeli Cyber War and Implications for the Future," *Military Review*, March–April 2003, http://usacac.leavenworth.army.mil/CAC/milreview/English/MarApr03/indexmarapr03.asp.

[13] Delio, Michelle, "Cyberwar? More Like Hot Air," Wirednews.com, May 4, 2001, http://www.wirednews.com/news/print/0,1294,43520,00.html.

[14] Gentile, Carmen, "Israeli Hackers Vow to Defend," Wired.com, November 15, 2000, http://www.wired.com/news/politics/0,1283,40187,00.html.

[15] Lemos, Robert, "Human Error Blamed for California Power Hack: Underscores Danger of Using Unsecure Networks," MSNBC.com, June 14, 2001, http://www.msnbc.com/news/587342.asp.

[16] Weisman, Robyn, "California Power Grid Hack Underscores Threat to U.S.," Newsfactor.com, June 13, 2001; http://www.newsfactor.com/perl/story/11220.html.

[17] Ackerman, Gwen, "Offering Refuge," originally on jpost, December 25, 2000, (Jerusalem Post Online Edition), http://www.jpost.com:80/Editions/2000/12/25/Digital/Digital.17923.html; reported on Infowar.com, December 26, 2000; http://www.infowar.com.

[18] Krebs, Brian, "FBI Arrests Hacker In Planned New Year's Eve Attack," *Newsbytes*, Washington, D.C., January 12, 2001, 4:54 PM CST; reported in Inforwar.com on January 15, 2001, http://www.infowar.com/hacker/01/hack_011501b_j.shtml.

[19] Krebs, Brian, "Feds Warn Of Concerted Hacker Attacks on New Year's Eve," originally reported in *Newsbytes*, Washington D.C., December 29, 2000; reported in Infowar.com, December 29, 2000, http://www.infowar.com/hacker/00/hack_122900a_j.shtml.

[20] James, Brig. Gen. Larry, "Integrating Space into the Leadership's DM Process," presentation at the *Space Core Technologies Conference*, November 6, 2003.

[21] "Information Warfare Tools Rolled Out in Iraq," *Jane's Defence Weekly*, August 6, 2003.

[22] Ratnezar, Romesh, "Hot on Saddam's Trail," August 3, 2003, http://www.time.com/timr/magazine/article/0,9171,1101030811-472819,00.html?cnn=yes.

[23] Chandrasekaran, Rajiv, "Car Bombs Kill at Least 35 in Baghdad," *Washington Post Foreign Service*, October 28, 2003, p. A01.

[24] "Veterans Back from Iraq with Stories to Tell," *The Washington Post*, March 19, 2006, http://www.msnbc.msn.com/id/11892317/.

[25] McGeary, Johanna, "Can the Iraqis Police Iraq?" *Time*, November 10, 2003, p. 45.

[26] Knickmeyer, Ellen, and K. I. Ibrahim, "Bombing Shatters Mosque in Iraq," *Washington Post Foreign Service*, February 23, 2006, p. A01, http://www.washingtonpost.com/wp-dyn/content/article/2006/02/22/AR2006022200454.html.

[27] Ignatius, David, "Fighting Smarter in Iraq," *The Washington Post*, March 17, 2006, p. A19, http://www.washingtonpost.com/wp-dyn/content/article/2006/03/16/AR2006031601308.html.

[28] CNN, "Rice Defends U.S. on Al Jazeera Network," CNN Online, October 15, 2001; http://archives.cnn.com/2001/US/10/15/ret.rice.aljazeera/.

[29] "Gestures, Mannerisms, Taboos," Iraq CD for pre-deployment preparation, Air Force Office of Special Investigations, http://public.afosi.amc.af.mil/deployment_stress/iraq/culture-gestures.html.

[30] Karon, Tony, "The Risks of an Iraq Election," Time.com, September 28, 2004, http://www.time.com/time/world/article/0,8599,702949,00.html?cnn=yes.

[31] Anderson, Jon Lee, "A Man of the Shadows: Can Iyad Allawi Hold Iraq Together?" *The New Yorker*, January 24–31, 2005; February 8, 2005, http://66.102.7.104/search?q=cache:64I2ITy6uCsJ:newyorker.com/printable/%3Ffact/050124fa_fact1+Iyad+Alawi+%22left+hand%22&hl=en.

[32] Clausewitz, Carl von, "Clausewitz Quotes," http://www.military-quotes.com/.

[33] Kessler, Glenn, and Robin Wright, "Bush's Words on Liberty Don't Mesh with Policies," *Washington Post*, January 21, 2005, p. A25, http://www.washingtonpost.com/wp-dyn/articles/A24581-2005Jan20.html.

[34] DNI, *Defense and the National Interest*, Kettle Creek Corporation, 2003, http://www.d-n-i.net/second_level/boyd_military.htm.

[35] Michel, Henri, *The Second World War*, translated by Douglas Parmee, New York: Praeger Publishers, 1975.

[36] Tomes, Robert R., "Boon or Bane? The Information Revolution and U.S. National Security," *Naval War College Press*, Summer 2000.

[37] Bonaparte, Napoleon, "Napoleon Quotes," http://www.military-quotes.com/.

[38] Catton, Brig. Gen. Jack, "Organizing for Influence," briefing for Infowarcon 2002.

[39] Allen, Patrick, and Chris Demchak, "An IO Conceptual Model and Applications framework," *Military Operations Research Journal*, Special Issue on IO/IW, Vol. 6, No. 2, 2001.

[40] Negron, Colonel Jose, "Analyzing Effects-Based Operations" white paper, Langley Air Force Base, VA: Air Combat Command, January 31, 2002.

[41] Rourke, Lt. Col. Bridget M., *Operational Net Assessment*, briefing USJFCOM J29, September 10, 2003.

[42] Snodgrass, Brig. Gen. Michael, "Effects Based Operations," *C4ISR Summit*, Danvers, MA, August 20, 2003.

2

The Increasing Importance of IO in Military Operations

> *Yes, I think we're vulnerable ... Some day someone will use those vulnerabilities to do great economic damage to the United States, or to slow and degrade our military response to a crisis.*
>
> —Richard Clarke

Information operations have been used throughout history. The manipulation of information is as old as warfare itself and likely as old as politics and society as well. However, the military dependence on information has also grown throughout history, so that modern militaries are now vulnerable to attacks against their information systems both on the battlefield and within their support structures. Now that we are in the Information Age, information and information operations in military conflict are more important than ever before. In addition to warfare, information is also now more essential than ever to the smooth running of civilian economies, societies, and governments. This leads to great opportunities for information-based economies, but also leads to great dependencies, and great dependencies lead to great vulnerabilities.

2.1 Early Examples of Information Operations

Elements of information operations have always been used in warfare. Some of the most often quoted examples of ancient use of IO are from Sun Tzu, who lived somewhere between 500 B.C. and 320 B.C. His maxims on war and politics are quoted in Joint Publication 3-13, the Air Force's Draft AFDD 2-5, and in Army War College textbooks. Sun Tzu provides several examples of information operations [1]:

- "All warfare is based on deception. Hence, when able to attack, we must seem unable; when using our forces, we must seem inactive; when we are near, we must make the enemy believe we are far away; when far away, we must make him believe we are near. Hold out baits to entice the enemy. Feign disorder, and crush him."
- "If you know the enemy and know yourself, you need not fear the result of a hundred battles. If you know yourself but not the enemy, for every victory gained you will also suffer a defeat. If you know neither the enemy nor yourself, you will succumb in every battle."
- "To fight and conquer in all your battles is not supreme excellence; supreme excellence consists in breaking the enemy's resistance without fighting."
- "The enemy's spies who have come to spy on us must be sought out, tempted with bribes, led away and comfortably housed. Thus they will become double agents and available for our service. It is through the information brought by the double agent that we are able to acquire and employ local and inward spies. It is owing to his information, again, that we can cause the doomed spy to carry false tidings to the enemy."

These examples include military deception, information on enemy and friendly forces, influence over the enemy without fighting, and counterintelligence. Not bad for 2300-year-old advice!

Napoleon was also noted for his use of deception in his early military career. For example, in his Italian campaign, he used a series of feints, decoys, forced marches, and daring attacks to confound his enemies and end a 5-year war in only 7 months [2]. His most successful deception occurred just prior to his great victory at Austerlitz. Napoleon wanted the Russian and Austrian commanders to believe [2] the "illusion that his army was disorganized, inferior in number, short of supplies and fearful of attack. To promote the ruse, he sent a negotiator to the allied headquarters to appeal for an armistice." When envoys from the Allied camp came to visit the French camp that night to corroborate their perceptions, Napoleon had companies of men frantically digging defensive positions around the visit area, and others running around with torches in the distance to convey the idea that this activity was widespread. The Allied envoys went back and dutifully reported the apparent defensive preparations that they had observed to the Austrian and Russian commanders [3]. In the morning, Napoleon deployed his forces with a clearly weak right flank, enticing the Allies to attack there. After the Allies committed their main force against this intentionally weak flank, Napoleon attacked with his main force, which had been partially hidden by hills. He split the Allied army in the center and rolled up the Allied left flank. Napoleon's use of deception was his key to victory against superior numbers [2].

Vladimir Ilyich Lenin was another master of the use and misuse of information to achieve his goals. Probably his most famous quote is: "A lie told often enough becomes the truth." He considered controlling what the populace heard critical to controlling the masses [4]: "When one makes a Revolution, one cannot mark time; one must always go forward—or go back. He who now talks about the 'freedom of the press' goes backward, and halts our headlong course towards Socialism." Controlling the schools and what the young learned was also key to gaining and maintaining control of the masses [4]: "Give me four years to teach the children and the seed I have sown will never be uprooted." To a great degree, Lenin was operating in the Information Age while the rest of the world was still in the Industrial Age.

Psychological operations against civilian populations are not new, either, as shown by the first example from ancient times [5]: "The Assyrians were also diabolically cruel, they skinned their victims alive, cut off their hands, feet, noses, ears, eyes, pulled out tongues, made mounds of heads and many more atrocities to inspire terror in those who were demanded to pay tribute." In World War I, the German military's "terror" policy was intended to ensure that the recently conquered populace would not cause them problems (which led to the Bryce report of German atrocities in Belgium) [6]. The Nazis expanded such tactics into a terrible art form in World War II, strafing refugee columns and unleashing the SS and Gestapo on restive populations, not to mention the use of concentration camps [7]. Douhet's principles of air power were designed toward intimidating the civilian population so that it would force the government to sue for peace [8]. The North Vietnamese campaign of weakening U.S. popular support was successful against the United States [9]. Militant Islamic cells have trained at least two generations to hate America, thereby providing a basis to justify actions that are contrary to the Koran [10].

So if all of these types of information operations have been around for millennia, what has changed to make information operations so important now?

2.2 Military Conflict in the Information Age

Information and information systems are more important on the modern battlefield than ever before due to the increased *dispersion of forces* and the need to quickly coordinate a wide range of physically distributed capabilities to achieve desired effects.

The primary reason for this dispersion is the ever-increasing lethality of weapon systems. Military historian Trevor Dupuy noted that the lethality of the individual soldier has increased dramatically over time. From individual arm-powered swords and spears, lethality increased with the longbow and the crossbow, followed by gunpowder, huge artillery, air and naval munitions, nuclear weapons, and precision munitions. The modern U.S. infantry platoon of around 30 soldiers inherently has more lethality—the ability to kill more people in the same period of time—than a Roman cohort of 800 to 1,000 men [11].

According to Dupuy, as that lethality increased, the dispersion of the individual on the battlefield also increased. Failure to disperse forces in the face of heavy firepower meant incurring large numbers of casualties very quickly, as the British discovered in Flanders fields, where they suffered 60,000 casualties (including 19,000 fatalities) in a single afternoon [12]. This dispersion has occurred not only on the battlefield, but also at sea, leading to the U.S. Navy's cooperative engagement tactics.

Dispersion, in turn, has led to an ever-increasing reliance on remote communications to coordinate actions. One of the German World War II blitzkrieg innovations was to have a radio on every tank, or at least with each tank platoon leader. Conversely, Russian tank units of World War II had such poor communications that they frequently had to stage their armor hours in advance of their use [13]. In a similar manner, radios became essential components in aircraft. Fleets no longer sailed in line as at the Battle of Jutland, but coordinated target information and fires via communications links. Modern militaries are now so dispersed as to be almost completely reliant on their information systems. It is that reliance that leads to vulnerability to interference with one's communications and information systems by the opposition, as well as opportunities for one to do the same to theirs.

IO, then, is more important in military conflict than ever before primarily because almost every military activity relies on information obtained by remote means and handled via geographically distributed information systems. Never before have militaries been so reliant on their information systems and processes. The rapid liberation of Iraq in 2003 was an example not only of that reliance, but also of the benefits of exploiting that capability [14].

The Information Age has affected the world's militaries, not just on the battlefield, but also on their whole support infrastructure. In the past, militaries have tended to have their own separate communications systems because these were more dependable than civilian communications systems, especially under adverse conditions. Militaries also tended to have their own logistics structures, which were more dependable than civilian supply infrastructures. In fact, military research and development in these areas often led the way to some of the great advances of the Industrial Age.

This trend has been reversed in the Information Age. Widely available civilian information infrastructure and technologies are, in many cases, more robust and dependable than similar technologies developed specifically for the military. For example, the use of the worldwide Internet in modern militaries is so pervasive that it is difficult to conceive of supporting operations without it. The U.S. military has also developed its own version of the Internet for use at the unclassified level (nonclassified Internet protocol router network, or NIPRNET), and at least two more levels of classified intranets [Secret Internet Protocol Router Network (SIPRNET) and Joint Worldwide Information Communications System (JWICS)] for the military and intelligence communities to share information. While the SIPRNET and JWICS intranets are physically separated from the Internet, there is always concern that the separation procedures will be violated by inattentive personnel [15].

Commercial computer systems and networking technologies, especially wireless technologies, are also being used by more and more military and defense agencies, which in turn leads to increased vulnerabilities, particularly as many civilian information systems have traditionally had little incentive to develop secure information systems. For example, the Defense Information Systems Agency (DISA), responsible for much of the information security within the DoD, used commercial wireless networks to connect their security cameras—without any encryption. This error was discovered by an intrusion detection expert using a wireless receiver from his car parked across the street from DISA [16].

As commercial information technologies continue to advance, the military will adopt more and more civilian information technology. However, as the world's militaries, and especially the U.S. military, become increasingly dependent on these commercial information technologies, their ultimate security is increasingly tied to the security of the information systems in their nation's civilian sector.

2.3 Advantage, Dependence, and Vulnerability in Information Age Societies

The Internet and other technologies of the Information Age have wrought rapid changes in economies and societies around the globe. Cell phones, pagers, and PDAs have become common elements in our daily lives. Schoolchildren have access to the largest library the world has ever known, where information on almost any subject can be rapidly obtained. Teenagers chat through instant messaging and e-mails with friends they may never have physically met. Employees work with colleagues for years before actually meeting each other, if ever. Work, play, entertainment, business, and home life have all changed dramatically in the last decade or two.[1]

This social saturation with information devices brings a number of advantages. The combination of the Internet, ubiquitous communications, and rapid, inexpensive transportation of goods all contribute to the modern economy. The economy benefits from the ability of small-business sellers to connect with a widely dispersed but much larger potential customer audience. Large businesses benefit by the ability to coordinate activities across the globe at a fraction of the previous communications costs and time. In 1998, the Internet economy generated $300 billion in revenue and employed 1.2 million people [17]. Just 2 years later, in 2000, the Internet economy had generated approximately $830 billion in revenue and employed just over 3 million people [18]. That's more than a 250% increase in just 2 years, and the portion of the economy based on the Internet has continued to increase every year since. JupiterResearch projected that Internet advertising alone would grow 27%, to $10.7 billion, in 2005. According to the Pew Internet & American Life Project, U.S. Internet use grew from 1 in 5 in 1995 to 2 in 3 in 2005, while 58 million Americans sent e-mail each day in December 2004 and 35 million used the Web to get news [19].

Activities on the global information grid are not limited to those with peaceful intent, however. Hackers prey on the unwary Internet visitor and e-mail user, which dampen the growth and confidence in the Internet economy [19]: "According to the Federal Trade Commission, 1 in 25 adults was a victim of identity theft in 2003 and the number of people affected online continues to increase ... The FTC said slightly more than half of the fraud-related claims it received in 2004 were Internet related, and many of the deceptions involved individuals or companies that used e-mail or a Web site." The results of a recent poll show that 48% of all Americans avoid Internet shopping due to concerns that their personal information will be stolen. Moreover, 93% of the respondents considered spyware a serious problem, and 97% considered identity theft a serious problem, but only 28% believed the government was doing enough to protect them [20].

Hate groups can now reach many more like-minded people, since neither geographical proximity nor broadcast capability is a requirement for contact [21]. Terrorists use the Internet to communicate plans, declare successes, and retain the headline-level attention so necessary for their success and survival. (See Chapter 6.) Cell phone services that provide instant messaging via the Internet can provide global communications to much of the known world [22].

---

1. The Cyberspace Policy Research Group has a number of articles on the effect of modern information technologies on militaries and other national agencies. See http://www.cyprg.arizona.edu.

Terrorists who live in isolated regions have been known to use satellite-based cell phones [23]. Terrorists don't own space-based capabilities, but because the modern global economy now offers space-based communication capabilities to anyone at an affordable price, terrorists can exploit these capabilities and remain well within their operating budgets.

The ubiquitous nature of information in the modern world also means that any local event may have international repercussions. The various communications media, media services, and propaganda channels have covered the electromagnetic spectrum and the Internet, so that little occurs that is not quickly known in most of the world (except for countries with strictly government controlled media). As a result, every event, however tactical, can suddenly expand to strategic, international importance. For example, both the images of the Abu Ghraib scandal and the images of the U.S. flag on the toppling statue of Hussein sent shockwaves through the Islamic world [24]. Conversely, U.S. soldiers often chose not to fire back at mosques from which they were receiving fire *because* they understood the potential international implications of such actions [25].

Modern civilian infrastructures are more vulnerable than ever before simply because they take advantage of the interconnected nature of the Information Age infrastructure. Companies become dependent on their Web-based transactions, e-mails, chat capabilities, wireless networks, mobile phones, faxes, PDAs, next-day delivery, continuous tracking, and just plain "connectedness." Being dependent means that one is vulnerable to losing that which one depends upon. For example, in a recent study where volunteers were not allowed to use the Internet for just 2 weeks, subjects felt terribly cut off, lost, and frustrated, and substantially less capable of functioning at work, at home, and in society [26].

In addition to the example in Chapter 1 of the Chinese hackers accessing an electric power network, other hackers from Saudi Arabia, Indonesia, and Pakistan have been found by the FBI to be electronically "casing" U.S. sites associated with emergency telephone systems, electric generation and transmission, water storage and distribution, nuclear power plants, and gas facilities [27]:

> Some of the probes suggested planning for a conventional attack ... But others homed in on a class of digital devices that allow remote control of services such as fire dispatch and of equipment such as pipelines. More information about those devices—and how to program them—turned up on Al Qaeda computers seized this year, according to law enforcement and national security officials.

Based on browser log data from a captured al-Qaeda laptop, al-Qaeda operators sought out and studied sites that offer software and programming instructions for the distributed control systems (DCS) and supervisory control and data acquisition (SCADA) systems that run power, water, transportation, and communications. Such SCADA devices are used to remotely collect sensor data, such as voltage and temperature in electric transmission systems, control railway switches, open or close circuit breakers, and adjust valves in water, oil, gas, and sewage pipelines [27].

In 2000, a hacker took control of a SCADA network in Australia, causing the release of hundreds of thousands of gallons of sewage into parks, rivers, and even a hotel. The hacker, Vitek Boden, hacked into Queensland's Maroochy Shire wastewater system 46 times over two months using commercially available remote SCADA equipment located in his car [28]. The SCADA system in Arizona's Roosevelt Dam in 1998 was hacked into and controlled by a 12-year-old hacker, who fortunately took no action once he was inside. Had his intentions been more sinister, he could have released the floodgates and swamped downstream communities, including both Mesa and Tempe [29].

In 2002, researchers in Finland identified a vulnerability in the Internet's routing protocols that could have been used to disrupt air traffic control, civilian and military phone links, and other critical infrastructure systems. These types of attacks on the infrastructure could be used by themselves but are more likely to be used in coordination with a large physical attack, so that the comprehension of and response to the physical attack are delayed and disorganized. Ronald Dick, the director of the FBI's National Infrastructure Protection Center, stated [27]: "The event I fear most is a physical attack in conjunction with a successful cyber-attack on the responders' 911 system or on the power grid ... [where] first responders couldn't get there ... and water didn't flow, hospitals didn't have power."

If one expands this dependency example from individuals to a nation or even the whole global economy, the implications for losing that upon which we are dependent becomes very clear. While the United States has taken some steps toward protecting our national infrastructure, our enemies have already targeted elements of that structure.

For example, Osama bin Laden believes that the Western economy is fragile and will collapse like a house of cards if pushed hard enough. The World Trade Center attack was designed not just to declare war on the U.S. homeland, but a real attempt to collapse the Western economy. The computers that tracked stock market transactions were in the two World Trade Center towers. Had it not been for a mirror site across the street that retained all of the stock market data, the collapse of the two towers could have crippled the Western economy for much longer than the four-day moratorium on trades. (The four days were necessary to validate and copy the data on the computers in the other building, which was also damaged in the attacks.) While the United States did suffer a decrease in the stock market after 9/11, the U.S. economy was much more resilient than bin Laden anticipated.

While bin Laden may have underestimated the resilience of the U.S. economy, the attack was also successful in becoming a rallying point to attack the United States and the West in general. Thus, 9/11 was a terrorist information operation both in terms of what it targeted (the Western economy), and as an influence operation, in terms of becoming a rallying point for anyone who hated the United States.

The so-called "digital divide" between the industrialized nations exploiting the Information Age and the developing nations still struggling to become industrialized also creates an imbalance in terms of vulnerability. Nations that use more of the Information Age technologies to support their economies are much more vulnerable to information operations than nations that rely less on these technologies. Because the United States has pioneered many of the advantages of the Information Age, it has become both the primary beneficiary of those advantages and the nation most dependent on them. Most of the Western nations are following suit. The discussion of the Palestinian/Israeli cyberconflict in Chapter 1 demonstrated the consequences of this imbalance of dependency between two antagonists [29].

Many of the developing nations, however, are not as dependent on their information infrastructures. Afghanistan under the Taliban did not have much of an information infrastructure, so it was not very vulnerable to IO attacks [30]. In a similar manner, the total number of personal computers in Iraq in 2003 numbered around 15,000, and most of those were state-controlled or state-monitored [31]. Even so, the lack of information infrastructure is not of long-term economic benefit, even if less vulnerable to IO attack in time of war.

Overall, the Information Age is leading to new opportunities, which lead to new dependencies, which lead to new vulnerabilities. Due to the economic advantages of the Information Age, most nations are taking advantage of these great new opportunities. The advance is being led by the United States, closely followed by the other Western nations and industrialized Pacific Rim nations, and more slowly by the developing nations. At the same time, exploiting these new opportunities leads to large dependencies, and large dependencies lead to large vulnerabilities. While the United States is economically in front of most of the world, it is also the most vulnerable to attacks against its own information infrastructure, as well as attacks against the world's information infrastructure. The U.S. and other Western militaries are also more dependent on their civilian information infrastructures than ever before. These are the reasons why IO has become so important not only on the modern battlefield, but also in the military support operations, and even in the survival of nations.

References

[1] Tzu, Sun, "Sun Tzu Quotes," http://www.military-quotes.com/.

[2] Horward, Prof. Donald D., "Napoleon 101," *Armchair General*, July 2005.

[3] Dupuy, Trevor N., *The Battle of Austerlitz: Napoleon's Greatest Victory*, New York: Macmillan, 1968.

[4] Lenin, Vladimir Ilyich, "Lenin Quotes," http://www.brainyquote.com/quotes/authors/v/vladimir_ilyich_lenin.html.

[5] BHO, "Archeology of Ancient Assyria," http://www.bible-history.com/assyria_archaeology/archaeology_of_ancient_assyria_ancient_assyria.html, November 2005.

[6] Halsey, Francis Whiting, *The Literary Digest History of the World War*, New York: Funk and Wagnalls Company, 1919.

[7] Leckie, Robert, *The Wars of America*, Vol. 2, New York: Harper & Row Publishers, 1968.

[8] Kelly, Michael, "The Air Power Revolution," *The Atlantic Monthly*, April 2002, http://www.theatlantic.com/issues/2002/04/kelly.htm.

[9] McCullogh, Declan, and Ben Poken, "What Future War Looks Like," Wired.com, September 18, 2001, http://www.wired.com/news/print/0,1294,46915,00.html.

[10] Ringle, Ken, "The Nature and Nurture of a Fanatical Believer," *The Washington Post*, September 25, 2001, http://www.washington post.com/ac2/wp-dyn?pagename=article&node=&contented=A19957-2001Sep24&notFound=true.

[11] Dupuy, Col. Ret. Trevor N., *Numbers, Predictions, and War*, Indianapolis, IN: The Bobbs-Merrill Company, Inc., 1979.

[12] Dupuy, R. Ernest, and Trevor N Dupuy, *The Encyclopedia of Military History from 3500 BC to the Present*, 2nd ed., New York: Harper & Row Publishers, 1986.

[13] Sharp, Charles C., *Soviet Armor Tactics in World War II*, West Chester, OH: George Nafziger, 1998, p. 89.

[14] Cordesman, Anthony H., *The Lessons of the Iraq War, Summary Briefing*, Center for Strategic and International Studies, July 15, 2003, http://www.csis.org/features/iraq_instantlessons_exec.pdf.

[15] Ruppe, David, and Johnathan Dube, "'Love Bug' Bites U.S. Military," ABCnews.com, May 5, 2000, fbox.vt.edu:10021/S/stsmith8/Debate1.html.

[16] Brewin, Bob, "DOD IT Projects Come Under Fire: Wireless LAN Security Lapse at Defense Agency," *Computerworld*, May 20, 2002, http://www.computerworld.com/securitytopics/security/story/0,10801,71306,00.html.

[17] Digital Ripple, "Factoid," *Ripple Effect E-zine*, June 24, 1999, http://www.digitalripple.com/rippleJun2499.html.

[18] Internet Indicators, "Measuring the Internet Economy," University of Texas and Cisco Systems, January 2001, http://www.internetindicators.com.

[19] Almasy, Steve, "The Internet Transforms Modern Life," CNN.com, June 24, 2005, http://www.cnn.com/2005/TECH/internet/06/23/evolution.main/index.html.

[20] Sullivan, Bob, "Data Leaks Stunt E-Commerce, Survey Suggests," MSNBC.com, June 15, 2005, http://www.msnbc.msn.com/id/8219161/.

[21] Lewis, George, "Hate Groups Multiply Online," MSNBC.com, March 2, 2005, http://www.msnbc.msn.com/id/7069514/.

[22] Straziuso, Jason, "Cell Phones Can Be Global Walkie-Talkies," CNN.com, September 19, 2003, http://cnn.netscape.cnn.com/news/story.jsp?oldflok=FF-APO-1700&idq=/ff/story/0001%2F20030919%2F105196551.htm&sc=1700&floc=NW_5-L7.

[23] CBS News, "Osama's Satellite Phone Switcheroo," CBSnews.com, January 21, 2003, http://www.cbsnews.com/stories/2003/01/21/attack/main537258.shtml.

[24] Fox News Staff, "Arabs Shocked, Awed by Fall of Baghdad," Fox News.com, April 9, 2003; http://www.foxnews.com/story/0,2933,83704,00.html.

[25] *Time* correspondent, unnamed, embedded with 101st Airborne, e-mail on his observations during his time in Iraq, 2003.

[26] Conifer Research, "Disconnected: Thirteen Families. Two Weeks. Zero Internet," 2004, http://promotions.yahoo.com/disconnected/.

[27] Gellman, Barton, "Cyber-Attacks by Al Qaeda Feared," *Washington Post*, June 27, 2002, p. A01, http://www.washingtonpost.com/ac2/wp-dyn/A50765-2002Jun26.

[28] Datz, Todd, "Out of Control," IDG Web site, August 12, 2004, http://cio.co.nz/cio.nsf/0/DA80828CFAE77CCACC256EED00752CB6?OpenDocument&More=Infrastructure+Feature).

[29] Associated Press, "Cyberwar Also Rages in Mideast," Wired.com, October 26, 2000, http://www.wired.com/news/print/0,1294,39766,00.html.

[30] Threat Analysis, "Al-Qaeda Cyber Capability," *Threat Analysis newsletter,* No. TA01-001, December 20, 2001, http://www.ocipep.gc.ca/opsprods/other/TA01-001_E.asp.

[31] "Information Warfare Tools Rolled Out in Iraq," *Jane's Defence Weekly*, August 6, 2003.

3

Characteristics of IO Planning

The complexity of operations in the information age will tax the ability of the ... staff to effectively handle all required tasks in a timely manner ...
—Concept Paper, United States Marine Corps

Information operations have features that make them unique compared to more traditional military operations. This chapter describes seven of the most significant characteristics that distinguish IO planning from traditional military planning:

1. IO involves a much larger *space of options* than traditional military operations, both in terms of the range of desired effects and in the number of IO capabilities potentially available.
2. The time and space dimensions of IO are *very broad.* The time dimension of IO can range from as small as milliseconds to as long as decades, while the physical space dimension can range from switching a computer "bit" to influence on a global scale.
3. Many IO desired effects cannot be achieved through direct action, but only through *indirect effects* resulting from a precise sequence of direct actions. In some cases, long chains of cause-and-effect sequences are required to achieve desired effects.
4. Due to this long cause-and-effect chain, *feedback* on interim results achieved and the application of "mid-course corrections" are of greater importance to the success of an IO mission than to many direct kinetic actions.

5. Because IO is both broad in scope and detailed in its function, a wide range of skills and expertise are required to successfully apply IO. These skills are often unavailable in combatant command headquarters. This leads to the need for *reachback support*, which leads to distributed planning, which contributes to the complexity of the IO planning process.
6. All of the preceding characteristics of IO contribute to a massive requirement for detailed information to form the basis of identifying and achieving the desired cause-and-effect chain. This dependence on large quantities of detailed information makes the *logistics of IO* very difficult to accomplish without its own efficient planning and preparation capability.
7. The fact that many IO capabilities are *highly classified* leads to a lack of exposure to the planners who are likely to be required to use them. This violates the maxim of "train as you fight" and precludes the ability to have a large pool of "veteran" IO warriors on the planning staff.

3.1 Feature 1: The Large IO Planning Space

The IO planning space is much larger than the traditional military planning space, which emphasizes the threatened or actual use of kinetic options (things that go boom) to achieve military objectives. Kinetic options are used to destroy people, equipment, and infrastructure; to threaten people into surrendering or retreating; to delay movement by damaging or destroying infrastructure; to deny movement across or ownership of terrain; and to degrade military and industrial activities. While traditional military operations have also sometimes employed military deception, psychological operations, operations security, and electronic warfare, these employments were used to support achievement of a traditional military objective.

Information operations are broader both in terms of the types of desired effects that can be achieved and in terms of the number of capabilities potentially available to achieve those effects. For example, the five desired effects of IO listed in the 1998 version of Joint Publication 3-13 were destroy, degrade, deny, disrupt, and delay. (These were also known as "the Five Ds.") In a study conducted in 2003, the author compiled a list of all the different desired effects from U.S. Joint and Service doctrine, IO concept papers, and the *Joint IO Handbook*, and found that a total of 44 desired effects were listed [1]. Table 3.1 presents this list in alphabetical order. Although in Chapter 5 we recommend a smaller set of desired effects, it is clear that the space of desired effects has far exceeded the original Five Ds.

Table 3.1
Sample IO Desired Effects Compiled from Various Sources

Access	Diminish	Mislead
Cascading network failure	Dislocate	Negate
Control	Disrupt	Neutralize
Coordination failure	Distract	Operational failure
Create information vacuum	Divert	Paralysis
Decapitate	Exploit	Penetrate
Deceive	Expose	Prevent
Decision Paralysis	Halt	Protect
Defeat	Harass	Read
Degrade	Influence	Safeguard
Delay	Inform	Shape
Deny	Interrupt	Shock
Destroy	Lose confidence in information	Stimulate
Desynchronize	Lose confidence in network	Stop
Deter	Manipulate	

This increase in the range of possible desired effects is due to the growing recognition that many options are available besides shooting something or threatening to shoot something. Since potential opponents are relying more heavily on information systems to manage their operations and capabilities, there are many more ways to foil an enemy plan, to force compliance by other than kinetic means, and to preclude a crisis from arising in the first place. For example, the enemy may want to know the status and location of our forces before they cross the border, and may hesitate to cross if they do not have this knowledge. Conversely, the enemy may not realize that the fuel they are depending upon to reach their advanced forces has been electronically rerouted to another location. IO planners seek to identify and employ a wide range of applicable desired effects to achieve national and operational objectives.

In addition to the large increase in the range of desired effects in IO, there has also been a similar increase in the number of potential capabilities. In general, any of the information dependencies described in Chapters 1 and 2 can potentially be exploited in some way. These capabilities could range from simple hacker techniques applied to gather military acquisition data [2] to an elaborate set of capabilities hypothesized in future-thinking books, such as a devastating cyber-attack on the U.S. stock market [3].

The combination of the increased range of desired effects that are potentially feasible in IO and the number of capabilities potentially available to achieve those effects leads to a very large option space for an IO planner to consider and coordinate. This large space of options requires a very broad and robust suite of planning techniques and tools to enable planners to adequately visualize and compare the IO options and tradeoffs space, as well as to select, execute, and monitor the resulting plans. Techniques and tools for visualizing this increased option space and integrating it into a coherent plan with both kinetic and nonkinetic options are addressed in Chapters 5 through 9.

3.2 Feature 2: The Time and Space Dimensions of IO

3.2.1 Temporal Scope

The time dimension of IO is very broad. Computer network operations involve software and switches acting in the realm of milliseconds, but can also involve the implanting of dormant software that could be launched days or months later. For example, Trojan Horse software, which allows the intruder repeated access at any point in the future, has been a standard technique of hackers for many years.[1] Conversely, psychological operations can last years and even generations.

In between these two extremes is a wide range of information operations that can span almost any planning horizon. The very breadth of time frames spanned by IO makes it essential to clarify the range of the time frames being considered in IO analysis, planning, and decision support. Therefore the *echelon* in which military plans are being prepared is the first important distinction to determine the planning horizon being considered. Moreover, planning tools need to be able to provide visualization techniques that allow decision makers and planners the ability to quickly grasp the scope of time being considered, and to be able to quickly change temporal scales to address critical issues in IO planning and execution. (See Chapter 9 for sample planning tools with broad and variable time scales.)

There are at least three more useful temporal distinctions applicable to IO. A second distinction is to distinguish between the time frame of the *action* and the time frame of the *desired effect*. For example, a leaflet drop often occurs within a few hours, while the desired effects of the leaflet may last for weeks [4]. Directly related to potential differences between taking the action and achieving the desired effect is the need to define *observation points* on the time line to ensure the action was taken and the effect achieved.

---

1. *Access at will* on a target system with a Trojan assumes that the Trojan is still active on the target machine and the target machine is remotely accessible in some way (such as via Internet or modem).

A third useful distinction is to specify the *phase* of an operation in which IO is being used. There are many categorizations of the phases of conflict, but the set of phases we will use here are peacetime, crisis, conflict, and postconflict. IO can and should be used in the peacetime phase to help monitor situations, influence the actors in that region, and help shape events so that conflict either does not occur, or occurs only in a form favorable to the United States. In the crisis phase, IO can be used to cause potential antagonists to stand down rather than start a shooting war, or again, if the shooting starts, to ensure that it is in a situation that is favorable to the United States. (Note that a number of IO capabilities could and should be used in peacetime and crisis phases of operation where kinetic options are not yet authorized. For example, influence operations were used against Iraqi decision-makers prior to the start of the shooting war in Operation Iraqi Freedom [5].) During the conflict phase, IO supports traditional military operations by providing both alternative and supporting options to the warfighter to achieve desired effects. In the postconflict phase, IO begins to resume its peacetime functions, but with an emphasis on ensuring that the previous conflict is actually over and in gathering the necessary information to support lessons learned and the updating of military planning factors. (See Chapter 8.) Moreover, influence-type information operations become the dominant mechanism to support stability operations and nation building. (See Chapters 6 and 9.)

A fourth useful distinction is to distinguish the time frame being planned for and *the time available to plan*. The time available to plan is generally a function of the phase one is in, along with situational conditions. In peacetime, deliberate planning is performed to prepare contingency plans for possible conflicts and to help influence world events and opinions to help preclude conflicts from arising. The amount of time available to perform deliberate planning can often be measured in months or years, such as supporting the biennial budget or the quadrennial defense review. During the crisis phase, the time available for planning reduces dramatically. Depending on the type of crisis, planning could be restricted to weeks or even days. (Some crisis response actions will be taken within hours of the event, particularly in the case of a natural disaster or terrorist attack.[2])

During the conflict phase, the planner and warfighter have much less time available to plan, and therefore consider fewer cases, examine each of these cases in less detail, and frequently rely on best guesses and assumptions in the absence of the opportunity to obtain more data. The next day's operations must be planned, or the plan refined, in a matter of hours. The next week's operations

---

2. Adaptive planning is being reenergized as being more responsive to changing circumstances and less reliant on the distinction between deliberate and crisis action planning [6].

will have days available to plan. Some of the next phase's planning may be underway throughout the crisis, and the time that portion of the phase is set to start determines the amount of time available to plan.

As the plan is executed, the space of decisions to be considered and analyzed is further reduced to variations on the current plan, such as branches and sequels.[3] (See Chapters 5 and 8 for more details on planning and employing branches and sequels in plans.)

Overall, the IO planner must consider the following temporal factors:

- Determine the *echelon* of the planning cell, which tends to define the time scale or planning horizon for the actions being planned.
- Determine the *phase* of the operation, which drives the time available to plan and the required level of detail of the plan.
- Define the *time available* to create, evaluate, compare, and select a plan for the selected planning horizon.
- Distinguish between the times *actions* are taken versus the time frames for the desired *effects* to be realized.
- Determine the timing of the *observation points* to provide the measurement of effect achievement, which is used for both mid-course correction of the executing plan, and to refine the planning factors over time. (Plan execution and planning factor feedback are described further in Chapter 8.)

3.2.2 Physical Scope

The physical scope of IO is also very broad. For example, computer network operations may focus at one time on the switching of one bit from on to off, and at another time on protecting the global grid. In a similar manner, psychological operations may focus at some times on the views of an individual, and at other times on national or transnational populations. An example of large-scale, long-term PSYOP by Islamic extremists is teaching generations of Muslims to hate Americans in particular and the West in general [7]. Conversely, individuals were targeted by some of the U.S. PSYOP against the Serbian leadership, such as Milosevic and his cronies [8]. In between these individuals and large

---

3. A branch defines a set of alternative actions, where one action will be taken if certain conditions are met, while a different action would be taken if different conditions occur. A sequel defines a single action that awaits a certain set of conditions to be met before being undertaken. Note that a branch may take option A, or option B, or await making that decision, while a sequel only has one option, and waits until the conditions for that option have been met before being activated.

populations are a wide range of mid-sized IO targets, not only in influence operations, but also in offensive and defensive IO. The scope of any type of information operation can range from the very small to the very large.

IO includes both tactical and global actions, but actions that are planned as tactical can also have global repercussions. As mentioned previously, the simple draping of the American flag over the toppling statue of Saddam Hussein had significant international impact in the Islamic world [9]. Similarly, the Abu Ghraib abuse photos were a significant boon for enemy propaganda and significantly reduced the probability of acceptance of subsequent U.S. messages [10]. These examples illustrate an encouraging trend toward understanding the political ramifications of very tactical events, and this trend of understanding is starting to permeate the military down to the individual soldier. The IO analyst, planner, and decision-maker must also be aware that some IO actions intended to be local or small in physical scope could have potentially global implications. This is one reason why IO military planning usually goes through a substantial interagency review process to ensure that what was intended to be tactical does not become unintentionally strategic. (See Chapter 5 regarding IO rules of engagement.)

3.3 Feature 3: Direct and Indirect Effects in IO

Another characteristic of IO is that its desired effects can frequently only be accomplished through indirect effects. Kinetic operations are usually considered direct effects because the cause-and-effect connection from action to desired effect is very close. For example, destroying an air defense missile site reduces the defender's ability to protect his airspace.

In contrast, information operations tend to have a longer sequence of links from direct actions to desired effect. While the ultimate objective of most IO is intended to affect the minds of the enemy decision-makers, IO operates against the information used to make those decisions and not against those minds directly. The opposing decision-maker bases his decisions on his perceptions, and his perceptions are based on available information. Thus, several steps are necessary from the direct action of, for example, attacking enemy bits and bytes, and causing the enemy commanders to make the decision you want him to make.

As a hypothetical example, assume we are going to try to cause the enemy commander to commit his strategic reserves to the southern front early in the campaign. We will employ military deception as the mechanism to achieve this early commitment. Part of the military deception involves penetrating the enemy computers used for military planning and decision support, and modifying key data to support the deception operation. The actions taken to map the enemy network, penetrate the network, and manipulate the data stored therein are all direct IO actions. That is, each step (map, penetrate, manipulate) has a direct action and a direct, measurable consequence. The achievement of each step can be estimated as a probability of accomplishment, and can be measured as to whether it succeeded, or the degree to which it succeeded, in changing the desired data. All of these are direct actions.

However, the ultimate objective is not just to change the data, but also to change the perception of the enemy commander, and further, to get the enemy commander to make the desired decision. Note that while the IO direct actions are objectively measurable, the next two steps—changing the commander's perception and getting the commander to make the decision—are one and two steps removed from any IO direct actions. Tweaking the data on the enemy computer is the direct action, but the change in perception and making of the decision are indirect effects of these direct IO actions. Due to the lack of the ability to directly affect the commander's perceptions and decisions, these steps are more difficult to control and the outcomes more difficult to predict. This has been compared to trying to control something behind you while all the knobs are in front of you [11].

Even the most perfect deception plan can fail simply because the opposing commander may make the correct decision for the wrong reason. Maybe the opposing commander is superstitious, or he has resources or people he wants to protect on the northern front. Maybe he just got out of the wrong side of the bed that morning. Since the opposing commander may still choose what we don't want him to choose in spite of a perfect deception plan, there are never any guarantees in IO that the desired effect will be accomplished. Any desired effect that requires unintended cooperation from the opponent can never have a 100% probability of success.

This lack of certainty makes planning for IO more difficult, since one is dealing with *probabilities of achieving objectives*, which may or may not lend themselves to confirmation by other means. Most military personnel will choose the relatively "sure thing" of destroying something than rely on a set of probable outcomes that may or may not achieve the desired effect. That is one reason why recent efforts at U.S. Strategic Command have focused on making IO planning steps as quantifiable and reliable as possible, so that the range of uncertainty is as small as the uncertainty associated with more traditional kinetic options [12].

3.4 Feature 4: The Role of Feedback in IO

A fourth characteristic of IO is that, due to these often long sequences of cause-and-effect chains to achieve the desired effect, there is a need to make sure that the chain is actually completed. If one were to look at the cause-and-effect chain as a "series system" in which every step must be accomplished, then one could simply multiply together the probability of achieving each step and determine the overall probability of success. However, this means that the longer the chain, the lower the probability of success, and the least likely step to succeed dramatically reduces the overall probability of success.

So how does one increase the probability of success in a long series system? The answer is to provide and act on *feedback* to determine whether each step has actually been accomplished, or if additional effort is required to achieve that step and therefore complete the chain. For example, in missile defense the principle of "shoot-look-shoot" is essentially feedback. If you missed the target the first time, you can fire a second time and still kill it. This is especially true if you know how much you missed by or why before you fire the second time. In any endeavor with uncertainty between direct action and ultimate objective, feedback provides the opportunity to get a second or third chance to get it right. The probability of success with feedback is always greater than or equal to the probability of success without feedback.

Timely and accurate feedback is essential to the success of most IO. Let us continue our military deception example described earlier, and further assume two cases. In the first case, all of the IO steps have to be defined ahead of time and then launched without feedback. In the second case, each step allows for feedback and a second or even a third attempt to accomplish the direct action.

In the first case, the plan is to map the network, penetrate the system, manipulate the data, change the perception of the commander, and obtain the desired commitment decision in the required time frame. The IO plan is executed, but somewhere along the way, something goes wrong and the plan fails. The strategic reserve is held in reserve, thereby remaining a threat to friendly military operations.

In the second case, we obtain feedback at each step of the way. We map the enemy network, only to find that there is a honey pot—a dummy portion of the network designed to attract and trap intruders. Through early feedback and further mapping, packet sniffing, and pattern matching, the honey pot is avoided and the target portion of the target network is identified. The penetration effort is then launched, but defensive computer security activity is detected. Therefore, the attack is shifted to another port, using a lower packet rate to stay hidden in the noise. Third, the desired data is located and manipulated as planned. However, we also discover an existing plan for relocating the strategic reserve further north. By tracking the supporting data for that plan, we manipulate more data to make that option less attractive. Finally, we monitor enemy plan development to ensure that the early commitment to the southern front is in the works, and provide additional data supporting the apparent urgency of committing those forces immediately. Note that feedback contributed to achieving the desired ultimate effect because the cause-and-effect chain was being monitored at every step of the way for degrees of success and the opportunity to improve the chance of the ultimate goal succeeding.

As the preceding hypothetical example illustrates, the IO planner can't just plan a long sequence of events and expect the desired effects to result. The distance from the direct IO action "control knobs" through the cause-and-effect chain to the final (indirect) desired effect is often too long to permit one to attempt to navigate that chain of events without feedback. IO requires a lot of tweaking and mid-course correction, making feedback essential to the success of IO.

Because of this, IO planning techniques and decision support tools must help the planner visualize the actual versus the planned, and the options available to correct any deviation or exploit new opportunities. Moreover, such a tool should have as contingencies preplanned responses to expected deviations so that the rapid response time in IO can be met and exploited. (See Chapters 5 and 8, which describe plan branches and sequels, which are mini-contingency plans embedded in a plan.)

3.4.1 Does He Know That I Know ... ?

Closely related to the issue of feedback is the threat of the *infinite feedback loop* dilemma in IO. Because IO is two-sided, a cat-and-mouse game is going on between the minds of the commanders on each side.[4] If I am trying to deceive the enemy to believe a deception plan, how do I know that he believes it? How do I know that he hasn't determined it is a deception, and is only pretending to believe the deception? Does he know that I know that he knows it is a deception? This does-he-know-that-I-know? cycle can go on ad infinitum, thereby making planning and execution monitoring very difficult to manage.

In reality, most decision-makers don't think at that many levels of perceptions of perceptions, because the combat situation and world events are often moving far too quickly for such feedback loops to carry significance for very long. In other words, the shelf life of feedback loops is very short and they deteriorate very quickly. For example, during the Arab-Israeli 1973 War, the Israelis decide to communicate in the clear on their tactical radios because it would take too long for the Syrians to translate the information and get it to their commanders in time to affect the battle [13].

It is usually sufficient for most operations to look at the first two or three levels of does-he-know and plan decisions based only on these levels. For example, look at what you know about the enemy, consider what it would mean to

---

4. IO is frequently multisided, but in this example, we are limiting ourselves to two sides for the sake of clarity.

your plan if he were to know that you have this knowledge, and prepare accordingly. It is not usually worth taking this feedback loop to four or five iterations unless there is some particular reason to expect significant results from extending the feedback loop. It is important to balance the marginal benefit of playing the does-he-know-that-I-know guessing game with the relatively near-term time frames of most planning horizons. Note that for endeavors with longer planning horizons, such as in the intelligence community during the Cold War, looking at multiple levels of feedback made sense because the situation was fairly stable and opportunities could be gained through small advantages in knowledge or perception. In a similar manner, the strategic conflict for influence (described in Chapter 11) could benefit from considering multiple levels of feedback. As before, the planning techniques and tools need to have the capability to allow planners and decision-makers to identify, visualize, and evaluate these feedback loops and their effects on the plan and its execution monitoring.

3.5 Feature 5: The Complexity of IO

The range of complexity is probably larger in IO than in any other military field, ranging from simple stimulus-response reactions to complex combinations of actions and counteractions.

IO is very complex because life is very complex. As mentioned earlier, tactical IO actions can have global implications. Moreover, efforts to cause an enemy decision-maker to take a desired action relies on IO direct actions to result in a desired indirect effect, as described earlier in this chapter.

In addition to the complexity inherent in long cause-and-effect chains and feedback loops as described above, the multisided nature of the *influence* arena of IO has an additional degree of complexity. It must be assumed that in today's globally connected information world, any message delivered to one intended audience will reach every audience, regardless of whether that degree of exposure is desired. There are many *groups of interest* that define an *idea battlespace,* an arena in which ideas compete for dominance—much like the movies *Tron* or *The Matrix* where computer programs battle each other for dominance in a cyber world. (See Chapter 6.) Messages interact with messages, messages interact with groups, and groups interact with groups. Moreover, the attention any group gives to a message will deteriorate over time unless the message is brought to mind through attention-getting actions, but repeating the same action too often will desensitize the target audience and attention will still wane.

Furthermore, messages do not usually fight on a fair playing field. Ethnic biases discredit otherwise valid messages; the identify of the messenger is important and subject to character assassination; messages can be subverted by associating them with fringe elements or just lying about them; and messages can be co-opted by the opposition as their own. This complex space of messages, groups, attention, and attacks on message content and ownership makes influence operations some of the most complex of IO. Modern conflicts and their eventual long-term outcomes are guided more by the media coverage and the perceptions that stay in the minds of the audience than by any military victory.

3.5.1 Organizing to Handle Complexity

The inherent complexity of IO implies that there is a need for a great deal of specialized knowledge and expertise across a wide range of fields. Due to the level of complexity of modern warfare and effects-based operations (including IO), a broad range of skills and knowledge is required to effectively support IO planning. However, the combatant commander (CC) staffs rarely have an expert in every field, whether civilian infrastructure, social networks, cyber security, or a wide range of nonkinetic options and interactions of interest. This means that the CC staff must be readily and consistently connected with such experts, or those sufficiently expert, to support plan development and decision making. Although *reachback* support to planning staffs is relatively new, it has shown promise in military operations over the last decade [14]. With new collaborative technologies and greater bandwidth than ever, the resources of many more centers of expertise are available to achieve the synergy among interactions and effects desired by EBO and IO.

These reachback centers and other agencies will have their own suite of decision support tools, including models, simulations, databases, and visualization tools to better understand the problem, propose solutions, explain their solutions to nonexperts, and document their solutions for review and audit. For example, the Joint Integrative Analysis and Planning Capability (JIAPC) will combine the resources of three agencies to provide holistic target descriptions useful to military planners [15]. Moreover, such organizations will be tasked by what this author calls RFAs—requests for analysis—which are similar to the requests for information (RFIs) used in the intelligence community. However, the RFAs include not only a target description, but also a more dynamic analysis of the potential desired and undesired effects on selected specific targets or categories of targets. Furthermore, the planner retains the audit trail of the supporting analysis to document how and why each element of the plan was selected for execution. (See Chapter 5.)

Due to the geographically distributed nature of reachback support, the planning techniques and tools will need to build in the ability to use and exploit reachback support as an integral part of the planning and execution monitoring process. These advanced techniques and tool suites will connect reachback centers and the CC staffs. This suite of tools will have a suite of interrelated and integrated software, database, and visualization techniques that are appropriate for supporting *distributed* deliberate, crisis action, persistent, and execution planning, as well as real-time or near-real-time execution monitoring, and a postanalysis process for reviewing and updating the planning factors just used.

3.6 Feature 6: The Information Requirements of IO

IO has a nearly insatiable requirement for information. IO planning often requires a huge amount of detailed information about each target, how it operates, how it responds to changes, and how to achieve desired effects without creating undesired effects. One of the reasons behind this is that both the range of desired effects and the capabilities potentially available are quite broad. For example, attacks against an enemy's infrastructure may require knowledge of electric power grids, transportation systems, oil and gas pipelines, and communications systems in general, and the infrastructure within the targeted nation in particular. This requires not only the substantial expertise required above, but also the detailed information about the specific infrastructure elements within the targeted nation.

To add further information requirements, a range of capabilities can affect each potential target, thereby increasing the combinations of information regarding how each capability can affect each target, the required conditions for the operation, and how it interacts with other capabilities are essential to IO planning.

A third reason for this requirement is the fact that the targeted decisionmaker's chain of dependence (outcomes depend on actions depend on decisions depend on perceptions depend on information) means that the attacker has to have some understanding of that dependence chain in order to be able to successfully affect it.

A fourth reason is that the IO attacker wants to achieve the desired effects and not potential undesired effects. Nations can be represented as a system of systems, and as such are highly interrelated. Achieving a desired effect in one part of the targeted system can also create undesired effects in other parts of the system. For example, knocking out electric power will force hospitals in the affected area onto generator backup, and if the targeted regime is heartless, they will cut off the supply of fuel to the hospital to claim innocents are being killed by the attacker. Therefore, detailed information is required to ensure that each recommended action will not result in undesired secondary effects.

As a result, the demand for information to support information operations can be huge. However, no one in the real world will ever know everything, so how does one plan for IO in the absence of perfect information? The answer lies in the fact that military personnel (and most business leaders) are trained to make decisions in the absence of information. The military maxim, "A good plan executed today is better than a perfect plan executed at some indefinite point in the future," applies to IO as well as military success [16]. So IO planning tools and methodologies simply need to do better than traditional techniques of providing the information required, and do not need to provide a complete set of perfect information.

3.6.1 Information Readiness

No IO planner will ever have all the information he or she would like to have, but the IO planner would like to minimize the risks by having enough of the right information. Much of this information will have to be collected during peacetime and refined during the preconflict phase, because access to the desired information once the shooting starts becomes even more difficult. *Information readiness* refers to having the required information available prior to when it will be needed. Much like the logistics of supplies and materiel, information readiness requires its own planning, preparation, and plan execution and monitoring. This problem of having the necessary information available is not unique to IO. For example, as the U.S. Air Force developed precision-guided munitions, it became apparent that these precision-guided munitions required detailed information not previously available in order to perform as designed [17]. So the logistics of IO is similar to previous information support problems, only much larger than ever before due to the increased demands of IO. The U.S. military has attempted to address some of these problems, and their proposed solutions are described in Chapter 10.

3.7 Feature 7: IO Classification and Protection Issues

The final characteristic of IO that we will discuss in this chapter is the fact that many of the real-world IO capabilities are often much more highly classified than equivalent traditional military capabilities. The primary reason is that many IO offensive and defensive capabilities could be fragile in that if the enemy knows how a certain capability actually operates, that capability could be readily countered by a piece of equipment, software, or operating technique. For example, during World War I, the Russians communicated on their radios in the open, and the Germans were able to exploit that to assist in the stunning victory at Tannenberg [18]. However, as soon as the Russians knew they were being listened to, they used codes to preclude simple eavesdropping. Dropping false leads when you think you are being listened in on is a common technique and used extensively during the Kosovo operation by the Serbs [19]. If a complex IO capability that takes time and money to research, develop, field, and maintain is readily compromised by simple steps once it is known to exist, then it behooves the owner of that capability to protect it under a heavy veil of secrecy. The good news about the classification of potential IO capabilities is that the risk of exposing a large number of IO capabilities all at once is very slim.

The bad news is that all of these classification restrictions lead to a large amount of ignorance within our own forces about our own capabilities, which leads to problems in the training environment, which leads to problems in the operational environment. This is more than just an inconvenience. If IO capabilities are not used in peacetime training, then they are often not used in actual operations [8, 20]. Since the U.S. military forces are supposed to train as they fight, this high degree of classification precludes the ability to practice in peacetime with the IO techniques that need to be incorporated into actual operations. Worse, due to lack of training with the techniques, the commander and staff are not veteran users of these capabilities and cannot effectively integrate them into Joint operations. It is difficult to provide enough qualified personnel to a combatant commander planning staff, let alone provide personnel with the proper clearances to know about the full spectrum of IO capabilities available. As a result, the United States either does not use the IO capabilities it may have in its arsenal or does not use them well—as part of an overall coordinated and fully integrated plan.

However, there are ways to mitigate these classification issues in the IO planning process, such as multilevel security processes and systems. For example, one can focus on the desired effect to be achieved against a specific target without specifying the method by which that desired effect is achieved. This reduces the risk to any of the technologies or methods that might be applied to achieve that desired effect. By keeping the details of an information operation at the appropriate level of classification, only the desired effect against the target is passed to the next lower level of classification. In a similar manner, the type of target affected also becomes more generic as the desired-effect/targeted-activity combination is further generalized as it passes to each lower level of classification. By the time the desired-effect/targeted-activity combination reaches the coalition planning level, the generalized combination can be maintained at the collateral secret level. Note that in order to pass the more generic information from higher to lower classification in the first place, some sort of ISSE Guard (Information Support Server Environment), Trusted Network Environment (TNE), or other multilevel security information passing system should be used.

Overall, these seven characteristics of IO define fairly unique requirements for planning methodologies and tools compared to more traditional military planning. Much of the emphasis will be on managing the complexity and breadth of scope inherent in IO, and packaging these characteristics in a manner that makes the problem manageable for the IO planner and the military and national decision-makers. These methodologies are described in Chapters 5 through 8, while some of the tools developed to support IO planning are described in Chapter 9.

References

[1] Allen, Patrick D., "Metrics and Course of Action Development for IO and EBO," Briefing for IWPC SPO, December 15, 2003.

[2] Bridis, Ted, "FBI Shorthanded in Battling Hackers," Associated Press, Web posted on *Augusta Chronicle*, October 8, 1999, http://chronicle.augusta.com/stories/100899/tec_124-8174.shtml.

[3] Clancy, Tom, *Debt of Honor*, New York: Putnam, 1994.

[4] Adams, James, *The Next World War*, New York: Simon and Schuster, 1998.

[5] French, Matthew, "DOD Aims Psyops at Iraqi Officers," FCW.com, March 24, 2003, http://www.fcw.com/fcw/articles/2003/0317/web-psyops-03-21-03.asp.

[6] "Adaptive Planning," *Inside the Pentagon* (ITP), April 15, 2005.

[7] Ringle, Ken, "The Nature and Nurture of a Fanatical Believer," *The Washington Post*, September 25, 2001, http://www.washington post.com/ac2/wp-dyn?pagename=article&node=&contented=A19957-2001Sep24&notFound=true.

[8] Arkin, William, and Robert Windrem, "The Other Kosovo War," MSNBC.com, August 29, 2001, http://www.msnbc.com/news/607032.asp.

[9] Fox News Staff, "Arabs Shocked, Awed by Fall of Baghdad," Fox News.com, April 9, 2003, http://www.foxnews.com/story/0,2933,83704,00.html.

[10] "New Abu Ghraib Abuse Images Revealed," MSNBC.com, February 15, 2006, http://www.msnbc.msn.com/id/11336276/.

[11] Allen, Patrick, and Chris Demchak, "An IO Conceptual Model and Applications Framework," *Military Operations Research Journal*, Special Issue on IO/IW, Vol.6 #2, 2001.

[12] Whiteman, Philip S., "A Tool for CNA COA Selection and Mission Analysis." *Proceedings of the 73rd Military Operations Research Society Symposium*, June 22–24, 2005.

[13] Richelson, Jeffrey T., *The U.S. Intelligence Community*, 2nd ed., New York: Ballinger, 1989.

[14] Donohue, Gen. William J., Congressional Testimony, March 8, 2000; http://www.fas.org/irp/congress/2000_hr/00-03-08donahue.htm.

[15] Henderson, Lt. Col. Garland S., "Joint Integrative Analysis and Planning Capability (JIAPC) Stakeholder IPT Briefing," Information Operations Directorate, OUSD/I(IO), July 14, 2003.

[16] Patton, George S., "Patton Quotes," http://www.military-quotes.com/.

[17] Hura, Myron, and Gary McLeod, *Intelligence Support and Mission Planning for Autonomous Precision-Guided Weapons: Implications for Intelligence Support Plan Development*, MR-230-AF, the Rand Corporation, 1993.

[18] Fowler, Will, "Battlefield Communications," Combat Online Web site, 2003, http://www.combat-online.com/radio.htm.

[19] Keys, Maj. Gen. Ron, "Future War and the Lessons of Kosovo," Briefing January 10, 2000.

[20] Thomas, Timothy, L., "Kosovo and the Current Myth of Information Superiority," *Parameters*, Spring 2000, http://www.carlisle.army.mil/usawc/parameters/00spring/contents.htm.

Part II
The IO Planning Process and Tools

4

Categories of IO Planning

A riot is a spontaneous outburst. A war is subject to advance planning.
—Richard M. Nixon

Chapter 1 described what IO is and is not and how IO planning must be considered as one element of effects-based planning that synchronizes both IO options with more traditional kinetic options. Chapter 2 described the historical and current importance of IO, while Chapter 3 described the issues unique to IO planning.

This chapter will describe the way in which U.S. doctrine has evolved over time, how DoD currently defines IO, and how these definitions map to the three categories of IO we will use in this book: attack, defense, and influence. Each of these categories has its own unique features with respect to planning and execution, and the distinctions between them are highlighted in this chapter. The details of IO planning methodologies for attack, influence, and defense will be presented in Chapters 5, 6, and 7, respectively, while execution monitoring and replanning are presented in Chapter 8.

4.1 The Evolving U.S. Doctrine of IO

U.S. IO doctrine has been changing quite a bit over the last few years. The second version of Joint Pub 3-13 was published on February 13, 2006 [1], the third version of AFDD-2-5 was published on January 11, 2005 [2], the third version of *Army Field Manual on IO* (formerly FM 100-6, now FM 3-13) was published on November 28, 2003 [3], and the second version of the *Joint IO Handbook* was published in July 2003 [4]. All of the Services and the

Department of Defense as a whole have been refining IO doctrine as their understanding of what IO is continues to evolve.

For example, the 1998 version of Joint doctrine (JP 3-13) offered the following definition of IO [5]: "Information Operations (IO) involve actions taken to affect adversary information and information systems while defending one's own information and information systems." The 2006 version of JP 3-13 defined IO as [1] "The integrated employment of electronic warfare (EW), computer network operations (CNO), psychological operations (PSYOP), military deception (MILDEC), and operations security (OPSEC), in concert with specified supporting and related capabilities, to influence, disrupt, corrupt, or usurp adversarial human and automated decision making while protecting our own."

The 1998 version of JP 3-13 also defined information warfare (IW) as [5]: "IO conducted during time of crisis or conflict (including war) to achieve or promote specific objectives over a specific adversary or adversaries." However, the JP 3-13 2006 version retires IW as a military term since the distinction between IO and IW was not well defined [1].

The U.S. Joint and Air Force doctrines were not consistent with each other, however. For example, the 1998 U.S. Air Force definition of IO was [6]: "actions taken to gain, exploit, defend, or attack information and information systems." Furthermore, this document defined IW as [6]: "information operations conducted to defend one's own information and information systems, or attacking and affecting an adversary's information and information systems." This version also included information-in-warfare (IIW)—which included things like position navigation, weather, communications, intelligence-surveillance-reconnaissance (ISR), information collection, and dissemination—as the second pillar of IO along with IW [6].

However, the 2005 version of AFDD 2-5 revised its definition of IO as follows [2]: "Information operations are the integrated employment of the capabilities of Influence Operations, Electronic Warfare Operations, Network Warfare Operations, in concert with specified integrated control enablers, to influence, disrupt, corrupt or usurp adversarial human and automated decision making while protecting our own." While this version dropped IIW from the realm of IO, it divided the space of IO capabilities differently than does Joint doctrine. For example, the 2005 Air Force version of AFDD 2-5 focuses on three IO operational elements: influence operations, electronic warfare operations, and network warfare operations—this differs from the five pillars of IO in the Joint doctrine (electronic warfare, computer network operations, military deception, PSYOP, and operational security) [2].

Meanwhile, the Army's FM 100-6 (1996 version) defined IO as [7]: "continuous military operations within the military information environment that enable, enhance, and protect the friendly force's ability to collect, process, and act on information to achieve an advantage across the full range of military operations; information operations include interacting with the global information environment and exploiting or denying an adversary's information and decision capabilities." However, the October 2002 draft of the Army's FM 3-13 gives the same definition of the core, supporting, and related elements of IO as the latest Joint doctrine [3].

Note that while the differences among the definitions of IO and its core capabilities vary somewhat among Joint and Service doctrines, the areas of agreement between these doctrine continues to grow with each iteration. This is a clear indication that our understanding of IO continues to evolve, but also appears to be settling down to a commonly agreed-upon set of definitions.

One reason for the differences between the Service definitions and the Joint definitions is the traditional issue of getting control of the budget over an area like IO. For example, if one used a kinetic weapon (like a precision guided munition) to destroy an enemy communications tower, was that an information operation or a traditional military operation? The answer depended on whom you asked, and whose budget was threatened. This problem was exacerbated by the fact that the original definition of IO was so broad that it made programming, budgeting, and organizational responsibilities for developing and managing IO capabilities very difficult.

A second reason for these doctrinal differences is that each Service culture is used to certain ways of thinking and organizing, which tends to funnel future thinking and concepts into old patterns. However, as the strength of the Joint community increases and the Service power structures start to realize the value in "Jointness," the differences between the Service definitions and the Joint definitions will continue to decrease.

Moreover, the DoD has been attempting different approaches for how it should organize itself to effectively and efficiently train, man, and equip IO capabilities. As a result, a number of major roles in IO have changed dramatically in the last few years. First, the IO mission was given to U.S. Space Command in October 1999 [8], but just a few years later was transferred to U.S. Strategic Command in July 2002 [9]. Pentagon turf wars have flared up not only among offices of different Military Services, but also among offices within a single Service. As more and more decision-makers are realizing the implications of living in the Information Age, they are attempting to adapt as quickly as large organizations can in the face of budget competition.

4.2 Three Useful Categories of IO

Clearly, something needed to be done to help the DoD organize and budget for IO as a distinct capability to achieve the goal of information superiority as defined in Joint Vision 2010. If the definition of IO is too narrow, it does not cover the real scope of IO. But if the definition of IO is too encompassing, decision-makers cannot plan, budget, and assign responsibilities properly. For example, Martin Libiki of the National Defense University stated that some definitions of IO were so broad as to "cover all human endeavor" [10]. His observation was quite accurate, in that some proponents of IO were pushing IO as encompassing everything. Although there is some rationale for making the definition of IO as large as possible, the downside is that it was bureaucratically unwieldy. When the definition of IO is too broad, then every organization can claim ownership of at least part of it, and turf wars became common. When the definition of IO is too narrow, the intended synergy between all aspects of IO is missing.

To address this problem of balancing adequate scope with fiscal and legal accountability or responsibility, the Office of the Secretary of Defense for Intelligence and Information Operations (OUSD(I) IO), proposed the following definition for IO in DoDD 3600.1: "... actions taken to influence, affect, or defend information systems, and decision-making" [11]. As the concept evolved, the IO Roadmap updated the definition of IO, and in November of 2005, it was officially released as a DoD definition in DoD Instruction 3600.2. That definition was also codified in Joint Publication 3-13 (2006), as [1].

> The integrated employment of electronic warfare (EW), computer network operations (CNO), psychological operations (PSYOP), military deception (MILDEC), and operations security (OPSEC), in concert with specified supporting and related capabilities, to influence, disrupt, corrupt, or usurp adversarial human and automated decision making while protecting our own.

These two definitions have two important features that were often present but not highlighted in previous definitions. First, the focus of IO is not just the information and information systems, but also the decision-making that these information systems support. The ultimate objective of IO is to influence decisions, and while affecting information systems is one way of doing so, there are others as well. Second, these definitions imply three broad categories for IO: influence, attack (affect or disrupt, corrupt, and usurp) and defend (while protecting our own). This is a useful categorization that we will build upon throughout the rest of the book, as shown in Figure 4.1. When performing IO planning, however, it is important to synchronize attack, defense, and influence actions as much as possible.

DoDD 3600.1 described the rationale behind which capabilities were core capabilities of IO, and which were related and supporting [12]:

> The Joint Staff viewpoint is that this is a means-based definition built around core military capabilities. It differentiates between IO and the informational element of national power, making the case that IO is not employed solely to affect adversaries. Information operations are key elements of national power, those which support effective national influence

Categories of Information Operations According to JP 3-13, Feb 2006			Mapped to Categories in this Book by Chapter			
			Attack	Defend	Influence	IO Logistics
Core Capabilities	CNO	CNA	Chap. 5			
		CNE	Chap. 5			
		CND		Chap. 7		
	EW	EA	Chap. 5			
		EP		Chap. 7		
		ES				Chap. 10
	OPSEC			Chap. 7		
	MilDec			Chap. 7		
	PSYOP				Chap. 6	
Related Capabilities	Public Affairs				Chaps 6, 11	
	Civil Military Ops				Chap. 6	
	Def Spt to Pub Dipl.				Chaps 6, 11	
Supporting Capabilities	IA			Chaps 1, 7		
	Physical Security			Chap. 7		
	Physical Attack		Chap. 5			
	Counterintelligence			Chap. 7		Chap. 10
	Combat Camera					Chap. 10
Other Support	Intelligence					Chap. 10

Figure 4.1 IO Categories According to JP 3-13 and Mapped to Book Chapters

that can preclude conflict or enhance national capabilities in warfare. In addition, Draft DoDD 3600.1 suggests that IO in the military should be considered in relation to core capabilities, related capabilities and supporting capabilities. All of these capabilities lie on a solid foundation of intelligence support. Information Operations are an important element of military power and national influence in peace, crisis, and conflict. Further, military information operations must support government policy, strategic goals, and desired end-states. Central to understanding IO is to understand that it is about influence, and effective influence requires interagency coordination and integrated regional engagement plans.

This new definition of IO capabilities is good for many reasons. First, it retains electronic warfare, computer network operations, military deception, psychological operations, and operational security as its five core capabilities. It relegates public affairs, civil military operations, and defense support to public diplomacy related capabilities, and information assurance, physical security, physical attack, combat camera and counterintelligence to supporting capabilities, while all of the above are supported by intelligence. This makes it clear, for example, that physical attack in support of IO is a supporting capability, but not a core IO capability. The benefit of this definition is that it helps define a clear distinction among the communities traditionally involved in different aspects of IO. When it comes to the Title 10 Service responsibilities of manning, equipping, and training for providing IO capabilities, this definition helps make those roles clear.

Second, it places into a supporting role the remaining fields of interest that were making IO too big to handle. Simply using information is no longer considered to be part of information warfare; otherwise, every commercial and human activity in the world would be considered an information operation.

Third, it allows the U.S. Military Services and the major commands within the Unified Command Plan the opportunity to divide up the space of IO responsibilities sufficiently to ensure that all of the important capabilities are covered, but are not more than the DoD can, or wants to, handle. It also helps the interagency process by helping distinguish the military contribution to IO from the State Department and National Intelligence contributions.

As mentioned in Chapter 1, the definition of IO is not just an academic exercise, because organizations build their justification for budget and power based on such definitions. Even when the Services do agree on the definition of IO, it will not necessarily end Service versus Joint competition in IO, or even competition within each Service.

4.3 Types of IO Operations Planning

The three main categories of IO (attack, influence, and defend) that we will use in this book have their own unique features with respect to planning operations in each realm. These distinctions will be addressed in Chapters 5, 6, and 7, respectively. This section will describe the different types of *operations planning* in which IO planning must be performed. Operations planning is performed to support specific military operations. Section 4.4 will describe some of the implications for organizing staffs to adequately support IO in operations planning. Section 4.5 will briefly discuss issues associated with fiscal or *acquisition planning* for IO.

The three types of "operations planning" currently performed by combatant commander staffs are deliberate planning, crisis action planning, and execution planning [1]. We will also discuss *persistent planning*, or adaptive planning, which is a relatively new concept associated with trying to make planning and execution more continuous and less cyclic. Further, we will discuss *execution monitoring* and replanning during operations, as well as real-time and near-real-time assessment and decision support. Finally, we will close this section with a discussion of the need for *postexecution assessment* and the feedback loop necessary to keep our planning factors up to date.

4.3.1 Deliberate Planning

Deliberate planning is what one does well in advance of a conflict—before being sure a specific conflict will actually arise. Deliberate planning supports nonacquisition long-term planning requirements, and is often called contingency planning. For example, the Rainbow series of plans prepared by the United States prior to World War II described actions to be taken in response to conflict with the Japanese. One plan, developed many years prior to December 7, 1941, recommended abandoning the Philippines until the United States could reconstitute sufficient force to defeat the Japanese [13].

Deliberate planning for IO involves long-term contingency planning associated with influence operations, defensive IO and civilian infrastructure defense, and offensive IO. In the influence arena, the United States had not even engaged in the battle of ideas for the minds of most of the Islamic world until September 11, 2001. Our enemies had at least two generations of creating mindless hatred and distrust of the United States across large sections of various populations. Since the United States is only recently beginning to respond to this situation, we have a lot of catching up to do, as described in Chapters 6 and 11. The time frame for deliberate planning to influence the minds of generations is very long indeed.

For military defensive IO and protecting civilian information infrastructure, deliberate planning involves setting policies, identifying weaknesses and critical nodes, and creating incentives for civilian organizations to improve their security. For example, many redundant communications links follow exactly the same path as existing links simply because it is cost-effective to do so. An incentive to move the redundant link to a physically separate location is much more secure, yet incurs a higher cost. Planning for how to identify and encourage more secure infrastructure decisions is important, even if it costs a little more, and such planning can cover a scope of many years.

For offensive IO, deliberate planning includes identifying vulnerabilities in potential opponents, planning for how to collect the necessary information to execute the plan, performing exercises to ensure things will go as expected, and gaining legal and operational review and approval.

The tools and techniques used to support deliberate planning can be extensive and detailed and take a long time to prepare, run, and analyze. The combination of breadth and depth is larger in deliberate planning tools than in most other planning and decision support tools because the time and resources tend to be available to perform such detailed analysis. Moreover, many more organizations are involved in deliberate planning to ensure that in-depth analysis is performed on critical assumptions and capabilities.

For all three categories of IO, the end product of deliberate planning is a contingency plan that may or may not ever be executed. These contingency plans are periodically updated, and are examined during crisis for applicability to the current mission and situation, as described next.

4.3.2 Crisis Action Planning

Crisis action planning begins when a warning is given by the national command authority that military action may be required in response to a given crisis. IO crisis action planning involves first determining if any appropriate IO plans exist for this crisis. If they do not, then new plans must be created from scratch. Even if plans do exist, it is likely that the information and assumptions upon which the contingency plans were created will need to be substantially updated.

In the influence category, the attitudes of groups of interest may have shifted over time, including their interactions and relative influence over other groups. In the defensive IO arena, the types and capabilities of the threats may have increased substantially, especially if some new flaw is found in commonly used routers and operating systems. In the offensive IO arena, the information infrastructure of our potential opponents will likely have changed, as could their defensive policies and procedures.

Unlike IO deliberate planning, IO crisis action planning has much less time available to be completed and evaluated. Requests for intelligence tend to be limited to those information items that can be obtained prior to the start of hostilities (which itself is an uncertain date). Moreover, the depth of analysis is limited by the time and resources available, which are much shorter and fewer than would be available for deliberate planning. The crisis action IO planning tools, therefore, must be able to be set up quickly, run quickly, and to have their outputs analyzed much more quickly than the tools used in deliberate planning.

Note that if the deliberate planning tools are available with appropriate data sets for the crisis at hand, they may be useful to the staff in modifying the contingency plan for this crisis action plan. However, the United States has frequently had to fight in places for which little or no material is previously available, such as Guadalcanal [14] and Grenada [15].

4.3.3 Execution Planning

Execution planning includes the planning that is *about to be* executed. Execution planning can occur before the shooting starts, or while preparing for the next phase of a plan being executed (like the next day's Effects Tasking Order). Execution planning has the shortest time frame for planning compared to deliberate and crisis action planning. In execution planning, the time when the plan will be executed is usually known and is usually measured in hours to a few days at most. Therefore, the tools required for execution planning must be even easier to prepare, faster running, and quicker to analyze than any of the preceding types of planning tools.

In the influence category, execution planning may involve the creation of a new message to counter an enemy message that has just appeared. For example, if an opponent lies and fabricates a story of a U.S.-committed atrocity, the proof that it is a lie needs to be created fairly quickly, approved, and disseminated, often within 24 hours. The same time scale is appropriate for creating a message to exploit an enemy error or atrocity. For longer influence operations measured in months, planning a shift on the ongoing influence campaign may be allowed as much as a week, but that tends to be the exception rather than the rule.

For defensive IO execution planning, much of the emphasis is on immediate shifts in the defensive policies and postures, as well as disseminating indications and warnings of what to look for preceding or during enemy IO action against our infrastructure. Relatively simple analysis tools of the effects of policy changes on the defensive posture are most often run at this time, because insufficient time exists to perform in-depth analysis.

Offensive IO execution planning involves defining what effects will be achieved by IO capabilities in this upcoming cycle of operations. Execution planners must perform a similar tradeoff between required responsiveness to meet the plan execution start time and the level of detail necessary to support each decision. Requests for information need to be very specific with very short desired response times.

For kinetic options, the development of the Joint Munitions Effectiveness Manual (JMEM) provides an estimate of expected damage caused by a given type of weapon against a target of specified hardness and an expected circular error probable (CEP). Although no such manual yet exists for nonkinetic capabilities, there is a need to develop such a JMEM-like set of planning factors for nonkinetic options, and U.S. Strategic Command has undertaken efforts to create one. Developing these execution planning factors for offensive IO will require experimentation and testing to make sure that these planning factors are sufficiently accurate to support good planning. In addition, a feedback process to evaluate and revise the accuracy of the planning factors will also be required.

4.3.4 Persistent Planning

Persistent planning was a concept developed by Lt. Col. Dennis Gilbert, U.S. Air Force, who noted that the need for information and assessment of the situation in IO actually needs to be ongoing and continuous rather than episodic. The shortened time frames in modern conflicts and the sheer magnitude of information required to support modern operations mean that there really is no downtime to planning anymore.[1]

Persistent planning is meant to be an ongoing endeavor that encompasses all of the traditional deliberate and crisis action planning, and even part of execution planning. Traditional deliberate planning was done periodically because the rate of change of military power was relatively slow. In the information domain, any potential battlefield is constantly changing. Whether it is a new

---

1. Adaptive planning being sponsored at the Joint Staff is a similar concept for continuous versus cycling planning and procedures for adaptive planning are being created [16].

operating system upgrade, a new communications channel, or a new message in the war for minds, any potential information battlefield in support of a military conflict will likely change more rapidly than the military domain and therefore requires much more constant vigilance. Persistent planning means that there is a continuous pipeline and feedback loop among requests for information, planning, assessment, and approval. Persistent planning is compatible with the concepts of both operational net assessment (or its follow-on) and predictive battlespace awareness, as described in Chapter 1.

4.3.5 Execution Monitoring and Decision Support

Execution monitoring and mid-execution decision support are used to ensure that the plan is executed as intended, to highlight and prepare for key decision points and to determine when deviations from the initial plan are sufficient to require replanning. Execution monitoring is different from execution planning, since execution planning is focused on planning how to execute the next mission. Execution monitoring collects information on how well the plan is going, assesses the degree of success in meeting objectives, identifies delays and problems, and determines what needs to be done next. Moreover, execution monitoring watches for planning monitors the key decision points, such as plan branches and sequels, and highlights the conditions that determine when one branch option should be selected or when a sequel should be launched. (See Chapters 5 and 8 for planning and monitoring branches and sequels.) For example, will a plan element's initiation time be delayed due to a detected enemy action? Or will a plan element be launched sooner than expected, such as the famous left hook during Operation Desert Storm [17]?

In influence operations, execution monitoring includes ways to directly and indirectly survey the results of friendly and enemy messages on selected groups of interest. Did the message achieve the desired effect? If not, what needs to be done to achieve that effect? The tools necessary to support influence operation execution monitoring, like other execution monitoring tools, need to be easy to set up, fast running, and with easy-to-understand outputs.

In defensive IO execution monitoring, some of the most common elements are automated trackers of indications and warnings. Commercial intrusion detection and monitoring tools are useful, but tend to have a high false-alarm rate. Other tools are required to corroborate information on attempted and successful intrusions. All of the information gathering, event assessment, and response analysis, planning, and approval have to occur in a very short period of time. This is especially true of automated response software designed to initiate a response in milliseconds. Such automated response mechanisms are not frequently used at this time due to the ease by which such automated responses can be spoofed. Since the decision to shut down a system is much more easily done by the owner and can do much more damage to the system performance than an external threat, most IO defensive monitoring systems are not designed to automatically make drastic changes to the system status without human intervention.

Offensive IO execution monitoring tools will likely have similar types of observation, assessment, decision support and approval components. Branches and sequels are mini-contingency plans that specify actions to be taken when certain conditions are met. This speeds the response time for key decisions and helps maintain the initiative by not spending substantial time at each key decision point.

4.3.6 Postcrisis Assessment and Feedback

After an operation or mission is complete, assessments should be made not only from the usual perspective of whether restrike is necessary, but also whether the planning factors were sufficiently accurate when comparing expected to actual results. Since decisions are made based on the planning factors, the accuracy of the planning factors drives whether the expected results closely matched or deviated widely from the actual results.

After World War II, the U.S. Army had a collection of planning factors in FM 101-10-1 (*Staff Officers' Field Manual: Organization, Technical, and Logistical Planning Factors*) that were used to guide industrial age warfare planning. Based on years of experience in a wide range of combat situations and environmental conditions, these planning factors were amazingly accurate. Even the planning factors for the amount of water consumed in a desert or arctic environment per man per day are still useful factors for logistics planning. As the nature of warfare has changed, however, the Army has retired these planning factor manuals, as operations are now conducted in the Information Age and often against unconventional threats.

The United States and our allies need comparable planning factors for the Information Age, and especially for planning and conducting information operations and other effects-based operations. The years of experience in performing such operations are not yet available to provide firm planning factors that planners can use with confidence. Even so, plans and decisions must be made in the absence of information, and planners and decision-makers will still use the best information available. What are needed in the near term are a process for creating these planning factors, a process for creating estimates on their uncertainty, a process to provide feedback to the keepers of the planning factors, and to decide whether these planning factors were accurate for the situation or if they need revision. Thus, any crisis or experiment could provide the opportunity to compare the accuracy of the planning factors and use them to update either their accuracy or to modify their uncertainty bounds. In particular, such feedback will help define which are similar and which are different situations so that the most appropriate planning factors can be used in each case.

4.4 Organizing for IO Operational Planning

To support the various types of IO planning described earlier, planners must be organized and equipped to provide the required level of analysis in the time available to perform each type of planning. Due to the range of types of IO plans, combatant commanders often divide up the work into various cells, where each cell performs one part of IO planning. For example, the IO planning cell is sometimes divided into offensive planning, defensive planning, and influence planning. They may be colocated or distributed, but if distributed, they often are well connected by communications channels to synchronize their plans. While it is true that to fully exploit IO capabilities, attack, defense, and influence should be planned by the same group, in reality, due to the wide range of skill mixes required to adequately address all aspects of IO, these planning tasks are often divided into separate cells or subcells within the IO cell.

The wide range of skill mixes required to adequately support all aspects of IO planning makes it very difficult to have an expert in every field, on every shift, in every headquarters. As discussed in Chapter 3, being able to access the experts remotely has been called *reachback* support. Although relatively new, reachback has shown promise in military operations over the last decade [17].

Different cells in the combatant command (CC) and their component staffs develop, monitor, and assess the progress of various portions of the plan within the planning and execution cycle. For example, Joint Staffs, with the assistance of component planning staffs, develop deliberate plans and iteratively collaborate with the other local and distributed planning staff members to refine those plans. As another example, crisis action planning is conducted by the Combined/Joint Air Operations Center (C/JAOC) Strategy Plans Team in support of the Joint Force Commander (JFC) and his staff, which also has the responsibility to plan for more than 96 hours into the future. The Strategy Guidance Team works on the 48- to 72-hour time scale to support execution planning and also monitors the progress of the current day's plan. The JFC and its components actively plan the next day's operations for execution and also monitor the current day's plan for achievement of objectives and upcoming key decision points. Meanwhile, the Operations Assessment Team provides feedback on what was and was not achieved in today's and yesterday's plan.[2]

---

2. While the terms Strategy Plans Team, Strategy Guidance Team, and Operations Assessment Team are Air Force terms, cells with similar functions and planning responsibilities but different names exist on the Joint and other Service staffs [18].

One reason for having different organizations perform deliberate versus crisis action and execution planning is that the tools available for, and required by, deliberate planning tend to be different than the other operations planning tools. In deliberate planning, the tools can be extensively detailed, can require lots of data, and can take a long time to set up, run, and evaluate. Conversely, crisis action planning and execution planning require a more specific, faster-running set of less-detailed tools to ensure that the planning time constraints are met. Similarly, execution monitoring tools tend to be very quick running and not very detailed so that they can meet the very short response times necessary to support branch and sequel decisions, as well as replanning decisions. There will rarely be time to run a complex analysis during execution monitoring.

4.5 IO Acquisition Planning

There is one more type of IO planning that is not part of operations planning: *acquisition planning*. Since the focus of this book is primarily on operations planning methodologies, we will touch only briefly on IO fiscal planning. (Chapter 10 presents some elements of acquisition planning with respect to training, experimentation, and testing.) One reason for the brevity of this description is that, at the time of this writing, there appear to be no methodologies or tools unique to IO to support fiscal decisions for the acquisition of IO capabilities.

Due to the comparative newness of IO to traditional military operations, there are as yet no models of IO comparable to TacWar, Thunder, ITEM, Combat Evaluation Model (CEM), or other combat models used to define requirements for conventional munitions and conventional system acquisition. Nor are there training models similar to the Corps Battle Simulation (CBS), Air Warfare Simulation (AWSIM), and the rest of the Joint Training Confederation. Moreover, claims that such traditional models adequately represent such considerations have been discounted for lack of credibility [19].

Although the lack of these analysis tools to support acquisition was identified years ago, the two simulation acquisition programs that were intended to address this shortfall have not yet succeeded. In the training realm, the Joint Simulation System (JSIMS) was severely over budget and underperforming when it was finally cancelled. In the analysis support realm, the new campaign-level Joint Warfighting Simulation (JWARS) program was also cancelled, although it was performing better than JSIMS. The remnants of the JWARS program are being transferred to and supported by U.S. Joint Forces

Command.[3] As a result, the analyst's toolset still lacks models to support fiscal decisions that include the unique features of IO.

Consequently, fiscal decisions for IO and IO-related programs still use the tried-and-true methods of decision analysis, spreadsheets, briefings, and reports. These techniques can usually be applied successfully to address the budgetary and programmatic issues associated with IO programs. There are just no good models yet to address the tradeoffs among alternatives in the land, sea, air, space, and information domains. Such a tool would be useful to support fiscal decisions and the tradeoffs among the various capabilities proposed for funding.

Once such models do exist, however, their representation of the strengths, weaknesses, and synergies among the various domains will need to be validated. For example, deep fires can be conducted and/or augmented by land, sea, air, and Special Forces assets. If the analytic tool does not represent all these capabilities, then the assets of one domain may be credited with unique capabilities that are not, in reality, unique. This need for adequate representation of strengths, weaknesses, and synergies is essential for tradeoff decisions and will have even a greater impact on decisions related to information operations assets. This is because analysts and decision-makers are less familiar with the actual capabilities and limitations of IO assets, and will thus be more likely to rely on the model results when their experience base is lacking.

In general, for fiscal planning support, *the level of detail required in the tools and analysis are directly proportional to the level of detail required of the decision*. If the decision is for broad-brush allocation of resources across various capability areas, then tools and techniques that apply to broad tradeoff categories will be appropriate. Conversely, if one is attempting to determine how much it will cost to design, develop, implement, and support a particular IO capability, then more detailed models of the specifics of that type of capability will be required to help support and justify the acquisition decision.

Overall, IO doctrine has been evolving fairly rapidly over the last decade, but the trend is toward increased agreement among the various definitions. Many of the remaining differences between Joint and Service IO doctrine are based on how the Services are choosing to organize themselves to man, equip, and train for IO. For the remainder of this book, we will focus on the U.S. Joint doctrine that defines IO with core capabilities of PSYOP, OPSEC, EW, CNO,

---

3. JWARS does contain the basics required to represent many IO features in the future because it bases its simulated decisions on perceptions rather than model truth, and tracks the passing of information in its communications model. However, the IO features in JWARS are still very rudimentary, and it will probably be some time before JWARS will be able to support acquisition planning and decision support for IO capabilities.

and MILDEC.[4] The types of planning we will describe include deliberate, crisis, execution, and persistent, as well as discuss execution monitoring of implemented plans and feedback on planning factor accuracy.

References

[1] Joint Publication 3-13, *Information Operations*, February 13, 2006.

[2] Air Force Doctrine Document 2-5, *Information Operations*, January 11, 2005.

[3] Field Manual 3-13, *Information Operations: Doctrine, Tactics, Techniques, and Procedures*, November 28, 2003.

[4] Joint Forces Staff College, *Joint Information Operations Planning Handbook*, July 2003.

[5] Joint Publication 3-13, *Joint Doctrine for Information Operations*, October 9, 1998.

[6] AFDD 2-5, *Information Operations*, August 5, 1998.

[7] Field Manual 100-6, *Information Operations*, August 27, 1996.

[8] Issler, Gordon D., "Space Warfare Meets Information Warfare," *Joint Forces Quarterly*, Autumn 2000.

[9] Rumsfeld, Donald, Secretary of Defense Memorandum, July 11, 2002.

[10] Libiki, Martin, "What Is Information Warfare?" Institute for National Strategic Studies, National Defense University, Washington, D.C., August 1995, http://www.fas.org/irp/wwwinfo.html.

[11] DoDD 3600.1, *Information Operations Policy*, December 2003.

[12] Deckro, Dick and Melissa Hathaway, "MORS Information Operations Workshop Summary," *Phalanx*, Vol. 35 No. 3, Military Operations Research Society, September 2002.

[13] Shaw, Henry I., Bernard C. Nulty, and Edwin T. Turnbladh, "Central Pacific Drive," *History of U.S. Marine Corps Operations in WWII*, online Hyperwar series from Historical Branch, G3 Division, Headquarters, U.S. Marine Corps, 2003, http://www.ibiblio.org/hyperwar/USMC/III/USMC-III-I-1.html.

[14] Leckie, Robert, *The Wars of America*, Vol. 2, New York: Harper & Row Publishers, 1968.

[15] FAS, "Operation Urgent Fury," Federation of American Scientists, 1999; http://www.fas.org/man/dod-101/ops/urgent_fury.htm.

[16] "Pentagon Overhauling Military Planning System for Future Conflicts," *Inside the Pentagon*, April 15, 2005.

---

4. Translating the Joint doctrine categories of IO to Air Force doctrine categories of IO is straightforward since the categories of IO are simply organized differently in Air Force doctrine.

[17] Schwarzkopf, General H. Norman, *It Doesn't Take a Hero*, New York: Bantam Books, 1998.

[18] *Air Force Operational Tactics, Techniques, and Procedures, 2-3.2*, December 14, 2004.

[19] Allen, Patrick, and Annette Ratzenberger, "Outputs Should Not Equal Inputs," *Phalanx*, September 2002.

5
Planning Methodologies and Technologies for Attack Missions in IO

> *He who stays on the defensive does not make war, he endures it.*
> —Field Marshal Wilhelm Leopold Colmar, Baron von der Goltz, 1883

This chapter describes the considerations and processes associated with planning offensive information operations, which include computer network attack and computer network exploitation from CNO, electronic attack from EW, and physical attack (see Figure 4.1). We describe offensive IO first, as it is easier in some respects than influence and defensive IO, and also because the latter use many of the same process elements and concepts as offensive IO. Thus, this chapter offers a good foundation for the discussion of influence and defensive IO in Chapters 6 and 7, respectively.

Many aspects are involved in planning offensive information operations. In this chapter, we will describe:

1. A methodology for how to create a Blue Capabilities Matrix by defining and visualizing the space of options available to the IO planner.

2. A categorization of the political constraints in which IO must operate and how the planner or commander can visualize the space of possible options within the limitations of those constraints.

3. A cause-and-effect network that will enable the planner or commander to visualize how desired effects support overall objectives and how each desired effect contributes to a change of state in its target that, in turn, contributes to achieving the desired end state.

4. The development and selection of a course-of-action (COA) analysis for *top-level* Blue-versus-Red comparisons and plan selection.
5. The development of COA comparisons for Blue *employment option* selection.
6. How these methodologies are designed to be scaleable to support both planning at different levels of support and reachback planning support. In this category, we describe the request-for-analysis process as a mechanism for the planner to request and track analytic support from reachback centers.
7. How these methodologies support planning among cells operating at different levels of security in a coalition environment and how this may result in a distributed plan.

5.1 Defining the Space of Options

Every commander and planner would like to know what options are available before making any plans or decisions. As mentioned in Chapter 3, the realm of the possible in IO is much larger than the more in traditional military operations. As a result, it is beneficial to explicitly describe to the commander and planners the space of options available to support operations in this planning period. We call this explicit description of the available options and their visualization the *Blue Capabilities Matrix* (BCM). This explicit definition not only makes planning more thorough and efficient, but it also helps provide part of the planning audit trail of what was selected, what was not selected, and why.

The first step in developing a BCM is to list the set of desired effects in IO, and then to define the types of enemy activities to which these effects can be applied. However, this is more easily said than done. As discussed in Chapter 3, the 44 desired effects shown in Table 3.1 (reproduced in Table 5.1) comprise the desired effects listed in various IO doctrine and planning guidance documents, such as JP 3-13 [1] and the *Joint IO Handbook* [2]. Upon closer examination, however, it becomes apparent that this list of desired effects is a mixture of apples and oranges.

For example, while some desired effects directly target information or information systems (e.g., access, destroy, penetrate), others target the *users* of that information (deceive, distract, influence), while still others refer to desired outcomes (lose confidence in information, operational failure). Some may be direct effects (e.g., hacking a network may provide access), while others are indirect secondary effects (discovering that a network has been hacked may cause a commander to lose confidence in information). Still others may be even longer-term indirect desired effects; for example, the hack (access) causes loss of

Table 5.1
Sample IO Desired Effects Compiled from Various Sources

Access	Diminish	Mislead
Cascading network failure	Dislocate	Negate
Control	Disrupt	Neutralize
Coordination failure	Distract	Operational failure
Create information vacuum	Divert	Paralysis
Decapitate	Exploit	Penetrate
Deceive	Expose	Prevent
Decision paralysis	Halt	Protect
Defeat	Harass	Read
Degrade	Influence	Safeguard
Delay	Inform	Shape
Deny	Interrupted	Shock
Destroy	Lose confidence in information	Stimulate
Desynchronize	Lose confidence in network	Stop
Deter	Manipulate	

confidence in information, which may in turn delay an enemy action, which might lead to an operational failure. Thus, while this table is a good starting point, more clarification is needed to distinguish between direct and indirect targets and effects, near-term versus long-term effects, and direct actionable effects versus long-term cumulative desired effects (which may not be directly actionable in terms of taking a specific IO action against a single target).

To address this problem, we developed Table 5.2 to distinguish between near-term and longer-term desired effects, as well as the types of targets against which each type of desired effect can operate. Note that this table is only for the list of IO attack options and does not cover the space of desired effects for defense or influence, nor for effects-based planning in general. The upper part of the table defines actionable or near-term desired effects, while the lower part defines longer-term desired effects. The columns represent the types of targets, including stored or moving information or data, the connectivity of the targeted information system, the information network itself, the functions or capabilities of an organization, and the ability of the nation or national authority to function.

For example, when the target is stored or moving information or data (as shown in the first target column), only four desired effects are applicable: destroy, delay, manipulate, and read-or-exploit. In the long term, these four actionable desired effects can lead to the longer-term desired effects of causing

Table 5.2
Distinguishing Desired Effects by Time Frame and Type Target

	Type of Target				
	Information or Data	Connectivity	System or Network	Organization Activity or Capability	Nation or National Authority
Near-term desired effects; direct effects; actionable	Destroy	Destroy	Destroy	Destroy	Destroy
	Delay	Disrupt or degrade capacity or performance	Disrupt or degrade capacity or performance	Delay	Delay decision
		Deny	Deny or isolate		
			Stop or shut down	Deter or stop	Deter
	Manipulate	Manipulate or control	Manipulate or control	Manipulate or control	
				Influence or shape attitudes	Influence or shape attitudes
			Stimulate	Stimulate	
	Read or exploit	Penetrate or compromise or access	Penetrate and access	Penetrate or read or exploit	
			Spoof; mislead	Distract or deceive or mislead	Distract or deceive or mislead
Long-term desired effects; indirect effects	Loss of confidence in information	Cease trying to connect	Loss of confidence in network	Paralyze or negate or neutralize	Decision paralysis
	Information vacuum	Block all connections	Cascading failure	Operational failure	
			Defeat	Defeat	

the enemy to lose confidence in his information, or presenting the enemy with an information vacuum.

With respect to attacking connectivity (the second target column), one can pursue additional desired effects such as disruption, denial, control, and penetration through direct action. The long-term result may be that the enemy may cease trying to connect, or be unable to connect because all connection ports or channels are blocked. Additional desired effects apply to the system or network level, organizational level, and national or national command authority level, as shown by the last three columns.

To *plan* such an attack, the planner would start by determining the desired effects on the targets at the national level, which in turn defines the desired effects on organizations, which determines the desired effects against the networks, connectivity, and/or information. When *executing* the plan, the sequence works in reverse, by seeking to obtain some sort of access through the enemy's connectivity, affecting the enemy's moving or stored information, possibly affecting the network, affecting the functions of targeted organizations, and possibly affecting selected functions of the targeted nation. Thus, the planning sequence starts at the right end of the table, while the execution sequence starts at the left (depending on the specifics of the plan).

This reduces the set of *actionable* desired effects for IO attack planning to the following (listed in alphabetical order):

- Degrade;
- Delay;
- Deny;
- Destroy;
- Disrupt;
- Exploit;
- Influence;
- Manipulate/control;
- Mislead or distract;
- Penetrate/access;
- Stimulate;
- Stop/deter.

The meaning of each of these desired effects is described in Appendix A, while the types of targets each can act upon is listed in Table 5.2. IO planners will undoubtedly choose to modify this list over time, and eventually doctrine will settle down and define the list that U.S planners will use. However, even if the list, or the meaning of each offensive IO desired effect, changes over time, the principle of categorizing desired effects by near term and longer term, as well as by the type of target against which each can act, is the important point to remember.

While Table 5.2 is useful, defining the Blue Capabilities Matrix still requires more elaboration on the types of targets or the specific functions of each target. Since the fourth column (organizational activities or capabilities) is fairly robust and generally applicable, we further refine the elements of an organization into a set of the following six target activity categories:

- Move;
- Shoot (to change a state—to perform primary function, such as repair);
- Sense;
- Communicate;
- Plan/think/decide;
- Protect/secure/store/retrieve.

Any organization will perform, or has the capability to perform, any of these six activities to accomplish its mission and attempt to survive. Since successfully "shooting" something is a means of changing the state of the target, we listed "shoot" as the descriptor for "change state," as it is easier to remember. Organizations that do not normally shoot (such as a repair depot) will still have a function that changes the state of something else (such as repair). Similarly, every organization senses, even if it uses only human eyes and ears. Every organization communicates, thinks, plans, and decides at some level. Every organization protects, secures, stores, and/or retrieves information and often physical objects (which may include the organization itself). Consequently, while this list may change over time, it has been found to be very robust for a wide range of effects-based planning scenarios, as these six functions scale across every echelon.

When we combine the list of actionable Blue desired effects with the targeted Red organizational activities, the result is one form of the Blue Capabilities Matrix for IO attack planning, as shown in Figure 5.1. The row labels are the Blue desired effects, while column headers are the targeted Red activities. Each cell in the matrix lists the number of "options" available to achieve the desired effect of that row against the targeted Red activity of that column. In a planning tool, clicking on the cell brings up the list of options available to achieve that desired effect against that targeted Red activity. Any given cell in the matrix may offer hundreds of options, or only a few, or none. In Figure 5.1, we color-coded (shaded here) the matrix based on the number of options available, but a better representation would be how effective the set of available options is against that targeted Red activity. Moreover, a planning tool representation of the BCM can also provide the planner with a drill-down capability to determine the details of that option, including its expected probability of success and possible or likely rule of engagement violations. (See Section 5.2 for satisfying or violating rules of engagement, and Section 5.5 for verifying plans to ensure that they remain within the political constraints.)

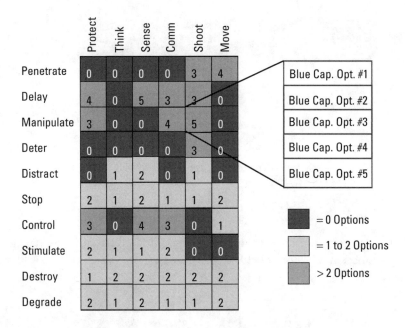

Figure 5.1 Blue Capabilities Matrix (BCM) and drill-down example.

Figure 5.1 is only one example of an IO attack Blue Capabilities Matrix. One could define such a matrix for a particular weapon system, so that the types of desired effects that weapon could achieve against a given targeted Red activity is known at a glance. Conversely, for acquisition planning, one could determine gaps in the set of capabilities to perform IO and help visualize and justify an acquisition because of the ability to affect some enemy activity not previously targetable. So the Blue Capabilities Matrix concept applies to a wide range of IO and effects-based applications. It is a useful visual and bookkeeping device that summarizes the space of options for a planner or decision maker.

Another example of a BCM has been defined for effects-based planning (EBP). In this case, the planner is looking at the instruments of national power [diplomatic, informational, military, and economic (DIME)] that could be applied to the targeted nation's capabilities [political, military, economic, social, information, and infrastructure (PMESII)].[1] Figure 5.2 shows the DIME-by-PMESII version of the BCM developed for the U.S. Joint Forces Command, and later applied in modified form to DARPA's Integrated Battle Command Program.

---

1. Although some scholars have argued for more instruments of national power besides the four comprising DIME, or to further subdivide DIME into more specific parts, or to add more factors to the PMESII list, the concept of the Blue Capabilities Matrix still applies regardless of the number of rows and columns defined for the BCM.

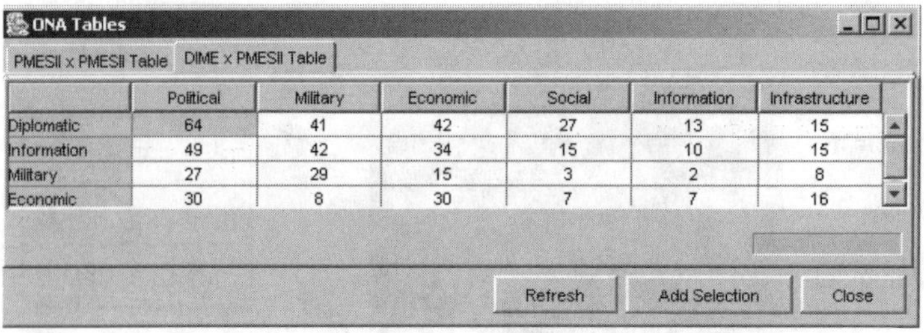

	Political	Military	Economic	Social	Information	Infrastructure
Diplomatic	64	41	42	27	13	15
Information	49	42	34	15	10	15
Military	27	29	15	3	2	8
Economic	30	8	30	7	7	16

Figure 5.2 DIME-by-PMESII version of the BCM.

Overall, the BCM helps the planner and decision-maker manage and grasp the large planning space of IO as described in Chapter 3. Moreover, distinguishing between near-term and longer-term desired effects helps manage the time and space dimensions of IO, and helps determine which desired effects are actionable.

5.2 Quantifying the Political Constraints

Modern U.S. military operations must achieve their objectives within a large set of political and other constraints. While most of our opponents, such as al-Qaeda, care nothing about the rules of war or the protection of noncombatants, the United States and most of our allies are rightly expected to achieve our objectives within a set of rules that constrain possible military behaviors [3].

Most planning tools to date have not provided planners with information regarding the constraints under which operations must be performed. These constraints are usually associated with a set of don'ts that define actions or end states to avoid while accomplishing a mission. For example, "Don't destroy the target nation's generators" is an example of a constraint upon the desired effect of disrupting a target nation's power generation and power distribution infrastructure. Thus far, planners have had to keep these constraints in mind, with no formal representation of the constraints or their potential achievement in the planning tools, or any audit trail of the decisions made (or their results) based upon such constraints.

To solve this problem, we came up with the following categorization of political and other constraints that can affect the planning process, and fully integrated these categories within an operational planning tool.

- Duration (maximum allowed duration for effect to persist);
- Extent—geographical (maximum allowed physical extent of the effect);
- Extent—functional (maximum allowed functional extent of the effect);

- Weapons authorization level (whether nonlethal or lethal conventional weapons have been authorized, or chemical, nuclear, or biological weapons have been authorized);
- Probability of collateral damage;
- Probability of secondary effects;
- Probability of detection (if we do it, will the other side know it?);
- Probability of attribution (if we use it, will the other side know we did it?);
- Proportionality composite (composite score of what is a proportional response);
- Loss or gain of intelligence (if we use it, do we lose or gain a source of intelligence?);
- Loss or gain of technology (if we use it, is the value of the technology lost, or is some technology gained?);
- Loss of allied business (what to avoid so we don't alienate our allies).

Note that this list is a list of the *categories* and does not yet define the specific constraints for each category, which will be a function of the mission and its political guidance. For example, the category of geographical extent might have a political constraint to preclude having the effects from a communication outage from extending into an adjacent allied or neutral nation. An example of a political constraint for the functional extent category might be to find a way to reduce military transportation without reducing all civilian transportation. Note that constraints regarding the categories of probability of detection and the probability of attribution will vary depending on whether or not conflict has started. For example, we may be greatly concerned about whether U.S. actions are detected or are attributable to the United States before the beginning of a conflict; after conflict has begun, however, we may no longer be concerned about whether the other side knows that we dropped a bomb. Even so, there may be some IO capabilities that we wish to keep hidden and will therefore always need to operate within an acceptable level of risk for detection and attribution.

The proportionality composite category represents a combination of factors that help determine whether a planned response is proportional to the original enemy action. For example, blowing up a whole port in response to Iranian mine-laying in the Gulf during the "flagging" operations was not considered a proportional act (and was avoided), but sinking one Iranian frigate and damaging another were considered proportional.[2]

---

2. In early 1988, Iran was attempting to interdict shipping in the Gulf as part of its war with Iraq. The United States responded with Operation Earnest Will to maintain freedom of

The loss of allied business category is becoming a more frequent consideration due to the globalization of the world economy. For example, during the Kosovo campaign, certain sites could not be bombed because this would cause a business loss to NATO allies who had commercial interests at the target site. Conversely, loss of enemy business is rarely of concern to the planner *unless* a nation is going to be occupied or liberated and the business will be essential to the future economy of that nation. Lastly, there is no category for gain of business because military actions aimed specifically at gaining a business advantage after the war is against the law of war and therefore automatically precluded.[3]

Note that the preceding list of rules of engagement (ROE) categories is a start and subject to modification over time. Different planning echelons will likely require additional or modified ROEs, and at the tactical level some of the ROEs might be weapon-specific. The point is that having a list of ROEs that is an integral part of the planning tools and part of each step of the planning process is an important new factor in IO planning, and military planning in general. Even though this is an initial list of ROE categories, this list has been reviewed by a number of legal and planning specialists and has remained fairly stable over time. As such, it acts as a useful starting point for IO and other effects-based planners. As the testing community has recently defined the representation of the satisfaction of these constraints as a Category I requirement, this list helps ensure that political considerations will be made an integral part of the planning process [6]. As with the previous categorizations, having the preceding list exactly right at this time is less important than using the concept to develop an agreed-upon standardized list for planning, visualization, and communications purposes.

Just as planners and decision-makers need to be able to quickly visualize the space of available options, they also need a way to be able to visualize the space of the constraints listed earlier and the effect of those constraints upon the option space during planning and execution. Our approach is to allow the higher command authority (HCA) and combatant commander to define a

---

navigation in the Persian Gulf. Iranian mining operations resulted in damage to the escorting U.S. frigate *Samuel B. Roberts*. The United States responded in force, resulting in the largest battle between surface forces since World War II. After one Iranian frigate, one gunboat, and several armed Iranian speedboats were sunk, the United States stopped the engagement when the second Iranian frigate *Sabalan* was dead in the water and burning. The United States allowed the Iranians to tow the crippled ship back to port to be saved and repaired [4].

3. For example, after the U.S. Civil War, the United States took Britain to court for assisting the Confederate States of America in building and outfitting commerce raiders, which devastated U.S. shipping interests and caused the United States to lose to Britain its market share of the global maritime shipping industry for decades thereafter. Britain was found guilty and paid $15.5 million in compensation [5].

standardized set of specific constraints within the standardized categories ROEs that will be used to guide subsequent planning. This top-down guidance will help preclude the problem of making detailed plans that will be rejected by the approving authorities. Such a capability provides the first quantitative and graphical representation of rules of engagement restrictions on military operations.

Due to the wide range of possible constraints for every possible action, we define a set of levels or bins that describe levels of behavior and control designed to meet the political constraints for each ROE defined in the above list of categories. For the duration ROE category, for example, we created the following four constraint bins (labeled A through D) to allow the HCA to select one for the duration category of the desired effect:

A. The effect lasts only as long as the action (e.g., jamming). The side performing the jamming action controls when the duration ends.

B. The effect lasts only as long as the primary effect remains in effect with no enemy action (e.g., mines reach lifespan and self-neutralize, non-persistent chemicals disperse). The side that delivers the capability knows when the duration will end, but cannot end it sooner.

C. The effect is permanent until *repaired* or neutralized by the enemy. The enemy must perform an action to recover full functionality, so the side that caused the effect cannot control when the duration ends, but can estimate when the enemy will repair that function.

D. The effect is permanent until *replaced* or neutralized by the enemy. This is similar to bin C except that the level of effect requires the enemy to replace rather than just repair the functionality, which usually takes longer. Again, the side that caused the effect cannot control when the duration ends.

A similar set of bins has been defined for each of the ROE categories so that the HCA can define which bins they desire the military to operate within for each ROE category. To provide further flexibility, the HCA can define both a set of ROE *guidelines* and a set of ROE *constraints* to guide the upcoming planning cycle. An ROE guideline defines the bin or level of each ROE category within which the HCA wants the planners to stay. An ROE constraint defines the same thing, except that a constraint is something that the HCA reviewer will approve only for compelling reasons. An ROE constraint is always greater than or equal to an ROE guideline. Figure 5.3 shows a sample set of ROE guidelines and constraints defined by the HCA. Note that guidelines (G) and constraints (C) can share the same bin, or constraints can be in a higher bin than guidelines.

G = Guideline
C = Constraint
C >= G

Levels for each category →	A	B	C	D
Duration		G C		
Extent (geographical)		G	C	
Extent (functional)		G C		
Weapon authorization			G C	
P(collateral damage)		G C		
P(secondary effects)	G	C		
P(detection)		G	C	
P(attribution)	G	C		
Proportionality composite		G C		
Loss or gain of intelligence				G C
Loss or gain of technology	G	C		
Loss of allied business	G	C		

Figure 5.3 ROE categories and ROE guidelines and constraints.

Having a standard set of ROE guidelines and constraints and a standard display mechanism provides planners with a quick and easy way to determine whether the review and approval authorities will consider that the ROEs have been violated. Planners may choose to exclude consideration of options that violate ROE constraints, since these are hard restrictions. Conversely, the planner may consider options that violate ROE constraints if there is sufficient cause to do so and attempt to justify such an action to the plan reviewing authorities. Note that for an option that violates an ROE constraint to be accepted, however, its measures of effectiveness (MOEs) should be significantly better than those for alternative options that do *not* violate ROE constraints. For example, it is unlikely that violating a more important ROE, such as a weapon level–authorized constraint against the use of nuclear weapons, would be accepted in a review even with substantial justification.

Note that even if all ROEs guidelines and constraints are met, a plan will still require review and approval. If, however, both the planner and the reviewing authority have an agreed-upon set of standard definitions and representations, the review process should take far less time than the current process, which lacks such standards. Such a list enables both the planner and the reviewing authority to identify quickly which, if any, ROE guidelines or constraints have been violated—and, thus, which items require more detailed rationale and justification. Such an approach supports the principle of management by exception. It also allows the HCA to define which types of IO are preapproved. (Until recently, the only preapproved IO capabilities have been in sonar and electronic warfare.) This, in turn, provides guidance to the combatant commander and his planning staff at all echelons as to the types of options and effects that will require review and approval.

In planning various options and COAs, the planner will expect certain outcomes to occur given the options selected. These expected results (R) can also be displayed relative to the guidelines and constraints, and color-coded for ease of reference. We use green for expected results that meet ROE guidelines and constraints, yellow for results that exceed guidelines but meet constraints, and red for results that exceed both guidelines and constraints, as shown in Figure 5.4. (In Figure 5.4, green is white, yellow is gray, and red is dark gray.)

Since each of the options in the BCM has an expected outcome with respect to each ROE, we can apply the ROE guidelines and constraints as "filters" on the BCM to display only those options that meet ROE constraints or meet ROE Guidelines and constraints. (See Figure 5.5.)

Note that as one looks from left to right, the number of options available to the combatant commander decreases. Since the guidelines are always less than or equal to the constraints, staying within the guideline filter is even more restrictive than staying within the constraint filter. In addition, this display capability allows the HCA to see, for the first time, a visual representation of the effect that the ROE guidelines and constraints are having on the operating forces. It also allows combatant commanders, and all planners at every echelon below them, to quickly see the space of options available. This process provides quick identification of the options that will be allowed under given ROE guidelines and constraints.

G = Guideline
C = Constraint
C >= G
R = Result

■ = Within guidelines
■ = Exceeds guidelines, but within constraints
■ = Exceeds constraints

Levels for each category	A	B	C	D
Duration		GCR		
Extent (geographical)		G	CR	
Extent (functional)		GCR		
Weapon authorization		R	GC	
P(collateral damage)		GCR		
P(secondary effects)	G	C	R	
P(detection)		G	CR	
P(attribution)	G	CR		
Proportionality composite		GCR		
Loss or gain of intelligence			GCR	
Loss or gain of technology	GR	C		
Loss of allied business	G	CR		

Figure 5.4 ROE expected results compared to guidelines and constraints.

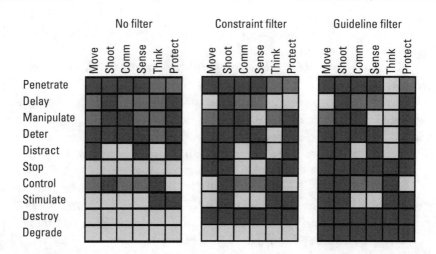

Figure 5.5 Blue Capabilities Matrix with ROE filters applied.

In addition to rules of engagement, planners of employment options must also take into account additional target-related considerations. For example, weaponeers and target planning personnel must select the specific friendly capabilities and actions to be used to achieve the desired effects against specific targets. In the past, when most capabilities were kinetic, it was sufficient to specify certain targets as no-strike targets. However, since the range of IO options involves much more than simply destroying a target, the planner must be much more specific in terms of defining what can and cannot be done against a specific target or class of targets.

We defined the following set of *target considerations* for planners to specify to ensure that the desired effect is achieved against a target or target type, while avoiding possible undesirable effects:

- No kinetic attack;
- No casualties;
- Electronic warfare operations only;
- Network operations only;
- Do not affect functionality X (specify in following text field);
- Only affect functionality Y (specify in following text field);
- No attribution to U.S. allies/coalition;
- No damage to civilian infrastructure;
- High value/high payoff target;
- NBC dispersion hazard (i.e., hitting this target may disperse NBC agents);

- Intelligence loss sensitive;
- Close proximity to sensitive site (like a religious or an archaeological site);
- Close proximity to friendly forces.

By specifying one or more of these target considerations before a plan is executed, the planner helps ensure that the weapons selectors and deliverers understand the considerations involved in selecting the target in the first place. As mentioned earlier, these factors should be definable for any specific target, or any category of target. For example, the Military Intelligence Data Base (MIDB) includes a set of target specification codes, from which we developed a subset of broader target-type categories to make the target consideration specification more efficient. Target-specific considerations override target-type considerations. In addition, the planner should be able to add new categories of target considerations to address local concerns; eventually these new criteria may be added to a standard list of target considerations.

Lastly, the planner also needs to define two time parameters: the minimum amount of time the desired effect is required to last, and the maximum amount of time the desired effect (or undesired consequences) is allowed to last. For example, cutting electric power to a radar site might be required to last a minimum of 30 minutes to enable a raid to penetrate enemy airspace, but as a maximum duration, the region's electronic power grid should not be shut down for more than two weeks.

The definition of rules of engagement and wider range of target considerations as integral parts of the IO planning process is a necessary and significant step forward, given the scope and complexity of IO planning. Moreover, for planners to be able to develop plans that will be approved and that will achieve the desired effects without serious undesired consequences, it is essential to have input as to what expected results are likely to exceed ROE guidelines and constraints. Lastly, maintaining an audit trail of these ROE expected outcomes and target considerations will ensure that more accurate planning factors are developed over time, since deviations between actual versus expected results will become apparent and spur the reevaluation of previous planning factors.

5.3 Cause-and-Effect Networks (CAENs)

Due to the complexity of IO, and the need to link direct and indirect effects to achieve the objectives, the planner needs a way to visualize the cause-and-effect relationships between the various parts of the plan. The cause-and-effect network (CAEN) is an extension of the U.S. Air Force's concept of strategy to task developed by Retired General Glen Kent in 1983 [7]. In the strategy-to-task concept, every action was required to be tied to some specific objective. For example, a bombing mission had to be performed in support of a specified objective, such as stopping an enemy offensive. Objectives were accomplished by tasks, and tasks were accomplished by actions. This was to preclude the problem that the U.S. Air Force faced by looking at missions as "servicing the target list." That is, the U.S. Air Force had a tendency to plan target strikes simply because a target was available, without defining *why* the target was being struck, or *whether* it even needed to be struck.

In the strategy-to-task hierarchy, each objective had one or more supporting tasks, and each task had one or more supporting actions. This resulted in a tree of objectives, tasks, and actions in which there was no cross communication or cross effects between actions taken on each branch of the tree. The assumption was that every action was independent of every other action, so planners compensated for any known cross connections by manually accounting for the timing of the various component parts of the plan on a synchronization matrix.

While strategy to task was a useful advance for traditional kinetic operations, the increased complexity of IO, and the interdependence between direct and indirect IO actions and effects, required a more robust representation of the causal relationships between various components of the plan. Cause-and-effect networks provide a way to visualize how the desired effects and their actions support the overall objectives and how each desired effect contributes to changes in state in the targets to help achieve that desired end state. As mentioned in Chapter 3, a long sequence of effects may be required to reach the desired end state. CAENs provide a mechanism for defining, tracking, visualizing, and communicating the sequence of effects and their actions. CAENs also provide the opportunity to explicitly address cross-connecting factors in a plan, thus enhancing the ability to better plan effects-based operations.

There are two primary distinctions between the strategy-to-task tree and the CAEN. The first is that the strategy-to-task tree is a tree, while the CAEN forms a network that shows the cross connections possible among the various elements. The second is that the CAEN includes a detailed representation of the adversary's dependencies, so that the planner can determine how changes in the state of one part of the Red capabilities may create problems for other parts of the Red capabilities. In other words, the cause-and-effect relationship continues beyond the Blue actions into the direct and subsequent (indirect) effects on Red capabilities, so that the complete cause-and-effect chain is explicitly defined. For example, it is not enough to say "target the enemy logistics" without some concept of how an attack on the enemy logistics will support the achievement of the final objectives.

Thus, a CAEN is a visual display of the whole plan (or part of a plan), including the Blue objectives to be achieved, the Red objectives to be prevented, the Blue (friendly) desired effects, and Red (adversary) activities. The CAEN shows not only the strategy-to-task relationship among all friendly actions, but also the interactions of friendly desired effects against targeted enemy activities. By defining the Red activities that must be accomplished so that Red can accomplish his objective, the planner can identify the types of Blue actions that will contribute to the achievement of the Blue objectives. This allows the planner to develop the plan by linking causes and effects in a logical chain, building on actions that achieve desired effects to accomplish the assigned mission. The Blue Planner may review the BCM to help guide the selection of Red activities to include in the Red activity dependencies.

Figure 5.6 shows an example of a CAEN in which the commander's guidance is to accomplish the mission by manipulating enemy decisions and plans, and by delaying and manipulating enemy supply flow. Note that the CAEN shows both the Blue objectives and tasks (boxes) and the Red mission and what it depends on (ovals).

The planner may obtain the Red dependencies from intelligence reports, predefined target descriptions, operational net assessment analyses, or previously prepared deliberate or crisis action plans. For example, if IPB analysis describes how the enemy plans to supply fuel to his forces via a pipeline, that information

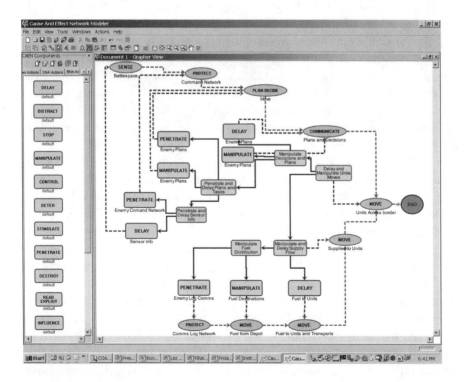

Figure 5.6 Sample CAEN.

can be inserted into the CAEN. If not already in CAEN format, the planner can add those nodes and links into the CAEN display. The planner can also determine whether any analyst products in CAEN format were produced previously during the deliberate planning or crisis action planning stages.

Various Blue Capabilities experts may also have already developed descriptions of Red activity trees that describe the processes that can be targeted by their weapon system. For example, a target developer or reachback center may have developed a model of a candidate Red Target process (such as logistics fuel flow or command and control network). The results of this analysis can be used to describe the dependencies of the Red activity on the target process, as well as help populate the MOE and ROE data for the Blue Capabilities Matrix.

Once the Red dependencies have been defined, the planner defines the subordinate tasks and desired effects to achieve the assigned mission. The planner selects from the list of desired effects and places them on the CAEN. The planner also selects Blue-to-Red links that connect the Blue desired effects to the targeted Red activities.

Two additional important planning features that cannot be easily represented in strategy-to-task trees are *branches* and *sequels*. A branch consists of two or more paths to take in the plan when appropriate conditions are met. For example, if the enemy commits his reserves to the North, we commit our reserves to the South, but if he commits his reserves to the South, we commit our reserves to the center. The default condition for a branch is wait until one of the path conditions is satisfied. Only one branch path will be followed, as soon as one of the condition sets is satisfied. Note that while COAs are developed, compared, and one selected during the planning process, branches are developed during planning, but they are assessed and one path is selected during execution, not during planning.

A sequel is like a branch in that it waits until the condition for its activation has been met. However, unlike a branch, there is only one path for a sequel. An example of a sequel is "When all units have reached phase-line X-ray, proceed to advance to phase-line Zebra." While sequels are defined during the planning process, they are only "triggered" during the execution process.

Branches provide the opportunity for planners to define embedded contingency plans (e.g., if this happens, do that), and to define key decision points for the commander. The focus on the conditions informs all decision-makers of a new command or commitment of forces when certain conditions have been met. For example, during Operation Desert Storm, General Schwarzkopf launched the left hook 48 hours early because the predefined conditions for its advance had been achieved early. If the left hook had not been launched ahead of schedule, a much smaller portion of the Iraqi Army would have been bagged in the encirclement [8].

A branch is represented in a CAEN as a node with two or more diverging paths, as shown by the circle in Figure 5.7. The conditions for the selection of the branch are stored with the branch node. In addition, the Execution Monitoring Tool (see Chapter 8) displays the key decision points on a timeline, and displays the status of the conditions at any point in time. A sequel is displayed on a CAEN on a single link, as the only question is when that action will be initiated, not whether it will be selected compared to an alternative.

The ability to represent the plan, including the why of each Blue action to support Blue objectives, as well as the subsequent chain of direct and indirect effects on enemy activities, provides a complete overview of the intended causal effects to be accomplished in the plan.[4] By explicitly defining, measuring, and tracking the Blue-to-Red causal links, the planner can manage the inherent complexity of IO and also help communicate these causal links to all planners, commanders, and reachback centers associated with the plan. Thus, CAENs will

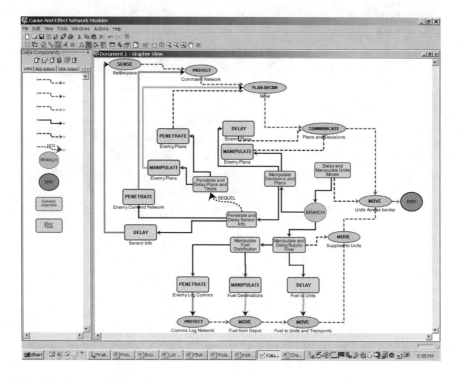

Figure 5.7 Sample CAEN with a branch and sequel.

---

4. The CAEN also allows the planner to insert unintended, undesirable, and secondary effects on any desired effect in the CAEN. These unintended effects can then be displayed or hidden by the planner.

be used as a central organizing concept throughout most of the rest of this book. Representations of the CAEN will be used to discuss how to plan influence and defensive operations as well, as described in Chapters 6 and 7, respectively.

5.4 Blue-Versus-Red COA Comparison

In this section, we distinguish between two types of courses of action (COAs). Much confusion has arisen when referring to COAs and tools that help plan and assess COAs, because not all COAs are designed to address the same types of issues. Just as the list of IO desired effects required some additional categorization and refinement to be useful, so do the definitions of COAs and COA development and comparison techniques.

The U.S. military's definition of a course of action is:

1. A plan that would accomplish, or is related to, the accomplishment of a mission.
2. The scheme adopted to accomplish a task or mission. It is a product of the Joint Operation Planning and Execution System concept development phase. The supported commander will include a recommended course of action in the commander's estimate. The recommended course of action will include the *concept of operations*, evaluation of *supportability estimates of supporting organizations*, and an integrated time-phased database of combat, combat support, and combat service support forces and sustainment [Joint Pub 1-02; emphasis added by the author]. Refinement of this database will be contingent on the time available for course of action development. When approved, the course of action becomes the basis for the development of an operation plan or operation order.

This definition hints at the two types of COA definitions that we will use in this book. The first type focuses on the *concept of operations* and more conceptual or top-level elements of warfighting. The second focuses on the supportability and consideration of various *employment options* to accomplish the higher COA. For lack of better terms, we will call the first type of COA the *top-level* or *conceptual* COA and the second type the *employment* or *supporting* COA.

The top-level (or conceptual) COA looks at the big picture, which compares the overall friendly and enemy options; examines broad-brush maneuvers and effects; and compares strategic, operational, or tactical advantages for each side (depending on the echelon doing the planning). Conceptual COAs also consider (either explicitly or implicitly) the possible courses of action of the enemy as part of their development. Whenever a planning echelon looks at the whole picture of the options before them and their adversary, they are creating a conceptual COA. For example, during Operation Desert Storm, the Coalition's COAs included the left hook across the desert, the up-the-middle play, and the amphibious "right hook" by the marines. This set of three COAs was developed, evaluated, and compared to determine which would best defeat the defending Iraqi forces. Developing higher COAs is definitely more of an art than a science.

The supporting or employment option COA focuses on how to best *accomplish* the given concept of operations or higher-level COA. There are many possible ways to achieve any desired effect. Once the desired effect is defined, the focus of the employment option COA is to determine how to best achieve that desired effect. Assume, for instance, that a given headquarters has developed a conceptual COA to a certain level of detail. The primary set of desired effects has been defined, but how to achieve those effects has not been defined. For example, if one desired effect is to disrupt the targeted nation's electric power, the supporting COAs to consider are whether it is better to target the electric distribution system or the electric generation system, and whether to use kinetic or nonkinetic options in either case. Thus, there are at least four employment COAs to consider in making the decision as to how to achieve the desired effect specified in the conceptual COA.

While the conceptual COA looks primarily at the "breadth" of options available to each side, the employment option COAs look at the "depth" of the various methods by which each selected conceptual COA can be accomplished. The conceptual COAs are much like a "game theory" type of problem, where the list of higher-level options available to each side is compared and assessed for probable relative outcomes. For these COAs, one is comparing possible Blue concepts of operation against Red concepts of operation so that Blue can select the best conceptual COA. The space of Blue and Red options is open-ended and not readily described by a set of well-defined options.

By contrast, the employment option COAs focus on comparing Blue options to achieve specified desired effects against targeted Red activities. The COA comparison method at this level is set within a fairly fixed or bounded problem space, and the set of options to be considered is usually well defined. For example, the entries in the Blue Capabilities Matrix define the space of applicable options to achieve desired effects that should be compared and selected for employment. Due to the relatively fixed option space, the employment option COAs lend themselves more readily to somewhat less subjective comparison techniques. At each stage of development, however, both COAs are constantly being refined, as feedback from the supporting COA may cause changes to the conceptual COA (e.g., when no employment options are found in the supporting COA to achieve the desired effects of the conceptual COA).

An example of examining employment option COAs follows. Since the Coalition decided to employ the left hook top-level COA during Operation Desert Storm, the Coalition needed to decide how to supply such a maneuver through a trackless wasteland. One employment option was to follow the traditional option to stockpile large quantities of fuel and fuel transports behind the front lines, which would later be driven forward to resupply rapidly advancing forces. Another employment option was to airlift substantial amounts of fuel to a remote airbase in the Western Desert well before the armored elements of the main Coalition force reached the area. Although the forward resupply base would be only lightly defended until the arrival of the main body, this option was chosen because it would better sustain the rapid advance, albeit with increased risk [9]. Note that for either supply option, the left hook was a given and the employment options considered were different ways to support the achievement of that left hook.

Note that whenever a subordinate headquarters examines its mission to achieve a desired effect, it is preparing a supporting COA for the higher-level COA. When the land component commander develops his COAs and selects one, the appropriate portion of the selected plan is passed to his supporting land component commanders for further development. This subordinate commander, in turn, is developing a supporting COA for achieving the mission assigned by the higher command, but is developing the top-level COA for his own command at the same time. Once he has selected his own COA, he sends that portion to his subordinate commanders, who perform the same sequence of top-level and supporting COAs for their echelon of command. Thus, one must remember that the higher command element is defining what is to be accomplished (the conceptual COA) while the subordinate commander is defining how to best implement it (the employment COA).

The remainder of this section will discuss development, comparison, and selection of the higher-level or conceptual COAs, and in particular the process of comparing Blue COAs against Red COAs.

5.4.1 Comparing Blue and Red COAs

The Joint Operations Planning and Execution System (JOPES) defines the need to compare Blue COAs to Red COAs before selecting the best Blue conceptual COA [10]. The JOPES process includes defining the most likely, most dangerous, and least likely Red COA (and possibly others), and comparing them to the two, three or more Blue COAs envisioned to accomplish the mission. This results in an N-by-M comparison, where N is the number of Red COAs being considered with respect to the M number of Blue COAs being considered. To keep the space of necessary comparisons to a reasonable amount, N and M usually do not exceed 2 or 3.

Note that *whenever* a Blue conceptual level COA is being evaluated, it is always with a Red COA in mind, even if the Red COA is not explicitly defined. In many cases, the Red COA is simply assumed. In our Operation Desert Storm example, the Coalition (Blue) left-hook, up-the-middle, or right-flank amphibious landing COAs each assumed an Iraqi (Red) force sitting in defensive positions. A more robust evaluation would have included the following three Red options: Defend in place, withdraw from Kuwait, and attack into Saudi Arabia. All three Red options had actually been defined and evaluated by the planning staffs early on, but as the preground-war situation unfolded, it became apparent that the most likely Red option was to defend in place. As a result, most of the subsequent plan refinement compared the Blue COAs against this single Red COA.

Each high-level or conceptual Blue COA can be represented by a cause-and-effect network, as can each of the Red COAs. At this point, the planner can benefit by a visual summary of the number and status of the Blue-versus-Red COA comparisons that must be undertaken. Figure 5.8 shows the nine combinations of three Blue COAs versus three Red COAs from Operation Desert Storm. We have annotated each Red COA as being most likely (defend in place), most dangerous (attack into Saudi Arabia), and least likely (withdraw from Kuwait).

At this stage, such comparisons are predominantly subjective in nature, as the level of uncertainty in these higher-level conceptual COAs is very large. As a

Legend

Score	Acceptable	Adequate
Complete	Distinguish	Feasible
Flexible	Suitable	Timely
Risk	Cost	

	Red COA 1: defend in place			Red COA 2: withdraw from Kuwait			Red COA 3: attack into Saudi Arabia		
	Most Likely			Least Likely			Most Dangerous		
Blue COA 1: Left hook across Western Desert	High	Yes	Yes	Med	Yes	Yes	Low	Yes	Yes
	Yes	Yes	Yes	Yes	Yes	Yes	Yes	Yes	Yes
	High	High	Yes	Med	Med	Yes	High	High	Yes
	Low	Low		Med	Low		Low	Low	
Blue COA 2: Up-the-middle through Wadi	Med	Yes	Yes	Med	Yes	Yes	Med	Yes	Yes
	Yes	Yes	Yes	Yes	Yes	Yes	Yes	Yes	Yes
	Low	Med	Yes	High	High	Yes	Low	Med	Yes
	High	Med		Low	Low		Med	Med	
Blue COA 3: amphibious landing from Persian Gulf	Low	Yes	Yes	Med	Yes	Yes	High	Yes	Yes
	Yes	Yes	Yes	Yes	Yes	Yes	Yes	Yes	Yes
	Med	High	Yes	Low	Med	Yes	High	High	Yes
	Med	High		Med	Low		Low	Med	

Figure 5.8 Blue-versus-Red COA comparison example.

result, modern planners mentally war-game, discuss, or otherwise subjectively evaluate the probability of how well each pair will achieve the desired results and avoid undesired results.

To facilitate a fair comparison, a criteria set is usually defined by the planning staff and often reviewed or approved by the commander. The same criteria set is used to perform each comparison. The weight of each criterion can be used to make each criterion more or less important to one another. Then, when comparing one Blue COA to one Red COA, the planning staff selects a subjective score (such as a numerical value associated with high, medium, and low) for each criterion. The total score for all criteria for each Blue-Red pairwise comparison is summed, and the result entered for that cell in Figure 5.8. The planners repeat the pairwise comparison process until all cells of the Blue-versus-Red COA comparison matrix have been filled in.

Once every Blue-Red COA comparison has been scored, the planners need to recommend one COA to the commander for execution. To select the best Blue COA, the planners and commander need to consider the variance in the expected outcomes, and the likelihood of each Red COA. For the variance consideration, one would prefer a COA whose scores do not vary widely versus each Red COA. For example, if Blue COA A does very well against Red COA #1, but very poorly against Red COA #2, and is mediocre against Red COA #3, then Blue COA A is not considered a robust solution. Conversely, if Blue COA B does well against Red COAs #1 and #2, and satisfactorily against Red COA #3, it has less variance and is considered a more robust solution. The desire for a robust solution must be weighed against the likelihood of the Red COA, however, since doing very well against the least likely Red COA does not carry as much weight as doing very well against the most likely Red COA.

It's important to note that the sources of these scores are still subjective, which means that as a whole, the Blue-versus-Red COA comparison is also very subjective. If sufficient time and resources are available, one could war-game each Blue-versus-Red COA comparison in a detailed combat simulation, but this, too, requires a great many assumptions and approximations, since information will be fairly sparse at this point in the planning. Even so, the top one or two Blue COAs are then selected for further refinement, and a number of employment option COAs are then developed for one or both to ensure that the achievement of the selected higher COA is actually feasible, as described in Section 5.5.

Since IO plan elements are embedded as part of the overall plan, no separate IO plan or IO COA is being developed. The IO elements of the overall plan are included as any other military capability. When employment option planning begins, more IO-specific options will then be considered, often in comparison to, or in combination with, other non-IO options. It is certainly feasible that the IO portion of the plan may become the dominant plan element, especially in the preconflict and postconflict phases of an operation.

5.5 Employment Options and the Blue Capabilities Matrix

Once a higher-level or conceptual COA has been defined, planners can begin to flesh out how each of the desired effects can be achieved, comparing employment options to determine the best option in each case. For example, a CAEN may seek to achieve dozens of desired effects. For each desired effect against a targeted Red activity, a number of options must be considered. The Blue Capabilities Matrix enables the planner to quickly look up the list of potentially applicable options that could or should be considered for any specific pairing of Blue desired effect and targeted Red activity. Figure 5.9 shows one way by which the options available in each situation can be presented to the planner.

Each option selection step for a Blue-to-Red pair (or Blue-to-Red link in the CAEN) can be considered a mini-COA analysis, where the benefits and weakness of each option can be reviewed and compared. For example, one option may have a higher probability of success, but may also include more

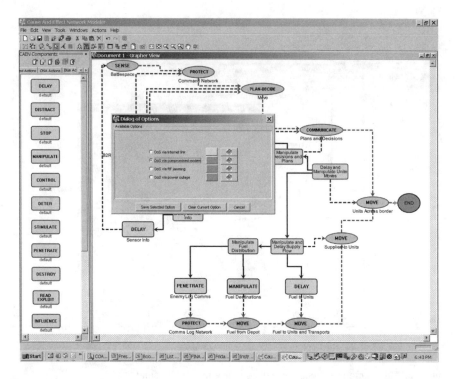

Figure 5.9 Selection of Blue capabilities for this desired effect and Red activity.

serious ROE violations. Figure 5.10 shows a sample display of the expected results of the selected option for a Blue desired effect/targeted Red activity pair.

The planner repeats this process of comparing and selecting options for each of the Blue-to-Red links on the CAEN. The result is a color-coded representation of the CAEN after all Blue-to-Red links have been populated with options. Each link is now color-coded to indicate whether *any* of the ROEs are violated (red [dark gray] for constraints violated, yellow [gray] for guidelines violated). If any ROEs are violated, that option (link) is tagged for further review.

The planner can now compare the various COA employment option variations that could support the higher-level COA that is being refined. One could simply look locally at each Blue-to-Red pair in the CAEN and attempt to select the best option independently. However, the problem of developing and comparing employment option COAs is not just a problem of selecting the "best" option on each Blue-to-Red link, but to select the best option *mix* for the overall COA, so one might also develop a set of COA variants that have their own employment themes. For example, one could develop a kinetic-heavy employment option COA based on the higher-level CAEN, or a nonkinetic-focused COA, or a mixed COA with a more balanced kinetic and nonkinetic mix. The

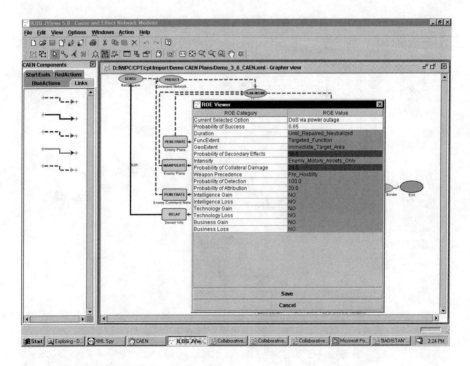

Figure 5.10 Expected results on ROE violations when option selected.

development of COA variants can help avoid the problem of selecting local optimum solutions that miss the natural synergy of certain option mixes.

Many different algorithms can be applied to the problem of employment COA scoring in order to compare COAs and help the planners select the best. However, all current algorithms have strengths and weaknesses, and none are completely satisfactory. Employment option COAs lend themselves to more quantitative techniques than do top-level COAs because the space of options is more clearly defined. Even so, the final decision for which employment option COA to select is still subjective and often based on factors that do *not* lend themselves to easy quantitative solutions. Moreover, relying too heavily on algorithms to make military decisions can create a level of predictability in friendly plans that is unhealthy in a conflict. (When your side becomes predictable, the enemy can take advantage of it.)

Figure 5.11 shows an example of how a COA comparison can appear when one considers both quantitative factors and whether each COA satisfies or violates ROE constraints. In this figure, Blue COA #1 appears to be the best choice because it has the highest probability of success is the highest and no ROE *constraint* violations. However, Blue COA #2 is very close to #1 in its

Variable Name	Course Of Action 1	Course Of Action 2	Course Of Action 3
COA Name	Main Objective COA A	Main Objective COA B	Main Objective COA C
COA Rank			
Score without ROEs	0.61	0.571	0.312
Score with ROEs	0.548	0.464	0.241
Probability of Success	0.291	0.338	0.317
Degree of Success	0.34	0.36	0.35
Confidence	0.9	0.9	0.9
Minimum Time	0.0	0.0	0.0
Maximum Time	0.0	0.0	0.0
Estimated Time	0.0	0.0	0.0
Duration	Until Repaired: Neutralized	Until Replaced: Neutralized	Until Replaced: Neutralized
Geo Extent	Within Target Nation	Within Target Nation	Immediate Target Area
Func Extent	Intentional Multifunction	Intentional Multifunction	Intentional Multifunction
Intensity	Enemy Military Assets Only	Enemy Military Assets Only	Enemy Military Assets Only
Weapon Precedence	Conventional	Conventional	Conventional
Probability of Detection	1.0	1.0	1.0
Probability of Attribution	1.0	1.0	1.0
Probability of Collateral Damage	0.3	0.2	0.7
Probability of Secondary Effects	0.25	0.15	0.5
Business Gain	Not expected	Not expected	Not expected
Business Loss	Not expected	Not expected	Not expected
Intelligence Gain	Not expected	Not expected	Not expected
Intelligence Loss	Not expected	Not expected	Not expected
Technology Gain	Not expected	Not expected	Not expected
Technology Loss	Not expected	Not expected	Not expected
Proportionality	Exceeds Guidelines	Exceeds Guidelines	Exceeds Constraints

Figure 5.11 COA comparison matrix for three COAs.

value for probability of success, and it has fewer ROE *guideline* violations. Therefore, the determination as to whether COA #1 or #2 is better must be evaluated subjectively, in terms of whether the political guideline violations are more serious for one than the other, given that their probability of success is within sufficient range of uncertainty to be considered equal.

A more difficult decision would be faced if COA #3 had a significantly higher probability of success, since it exceeds three political constraints. Since it violates the proportionality, secondary effects and collateral damage constraints, it would probably be rejected unless no other feasible COAs were available or the consequences of not selecting it were egregious. Once again, such decisions cannot be based on strictly algorithmic results, but must involve careful human evaluation of the complete picture. That is why it is important to rank the COAs in terms of best to worst, so that if, for any reason, the preferred COA is rejected, the second-best COA is still available for further refinement and selection.

Another advantage of explicitly representing the ROE violations for employment option COAs is that the resulting feedback to the conceptual COA may indicate that refinement of that COA, or selection of the next-best conceptual COA, would be preferable to selecting any of the current employment option COAs.

Whichever COA is selected for the plan to be executed, the assessment of the military, informational, and political factors can be stored in an electronic audit trail that documents which COA was selected, which was not, and why. Such an audit trail is essential in the modern world, where the United States has already been called upon to justify before the World Court to defend why certain military decisions were made [personal communication with Jeff Walker, October 7, 2003].

Overall, understanding the distinction between conceptual COAs and employment option COAs helps planners understand the boundaries of their decision space at each stage of COA development, evaluation, comparison, and selection. While employment option COAs lend themselves to more quantitative development and scoring techniques, the mind of the planner and commander is still essential to ensure the comprehensive understanding of the decision space, including political and informational as well as military aspects of planning. The IO planning methodology presented here helps the planner manage the large IO planning space, including both direct and indirect effects; causal links among Blue-to-Blue, Blue-to-Red, and Red-to-Red relationships; the inherent complexity of IO planning; and key elements of effects-based planning.

5.6 Scalability, Reachback, and Requests for Analysis

As mentioned in Section 5.4, once a higher or conceptual COA has been developed, it needs to be refined in order to determine its actual feasibility or desirability compared to other COAs. Planners further refine each COA through one or more of the following three methods:

- By considering employment options in the current CAEN(s) to determine how to employ assets to accomplish the plan objectives;
- By passing an appropriate section of the plan or COA to a subordinate command for further refinement by adding new, more detailed CAEN elements and/or inserting and evaluating employment options;
- By passing an appropriate section of the plan or COA to a reachback center, usually for elaboration of a set of employment options.

Note that a given plan may have one section being further refined by the original planning echelon, another part being elaborated upon by a subordinate command, and yet another part being evaluated by a reachback center. The ability to have a real-time collaborative planning capability is essential to the success of these parallel planning steps, as described further in Chapter 9.

The scalability of this process becomes apparent when considering the process steps at each echelon of command and the reachback centers. Each echelon performs planning that elaborates on the higher-level COA provided by its superior headquarters. For example, when any corps headquarters assigns a mission to a division headquarters, the division headquarters is responsible for developing its own COAs to ensure that the employment options will achieve the objectives of the higher-level COA. Each echelon of command performs the same set of planning steps, defining the higher-level COA for its echelon of command, and either passing it to one of its subordinate commands for further refinement, or sending it to a reachback center for employment option elaboration, or performing employment option analysis and selection at their echelon of command.

Moreover, the structure of the CAEN ensures that the subordinate commands know exactly where their new CAEN elements belong in the overall COA or plan. When the detailed CAENs are returned to the higher-level command, the higher-level command also knows exactly where these new elements fit into the plan. If the higher and subordinate commands share access to the same COA and plan database, then new subordinate elements become part of the overall plan as soon as they are defined, they are already part of the overall plan. If the two echelons of command do not share access to the same database, then some sort of marker points in the plan are used to ensure that the subordinate plan elements are placed into the proper locations in the higher echelon's CAEN. This is called a request for detailed planning (RFDP).

Employment option COAs can be created, evaluated, compared, and/or selected at the higher headquarters, the subordinate headquarters, or a reachback center. The higher headquarters planner may want to create an estimate for how well a given employment option would succeed, whether it would violate any rules of engagement, and what the likely set of target considerations would be for a broad set of target categories. Subsequently or simultaneously, a subordinate headquarters can be developing a more detailed CAEN with additional Blue desired effects and targeted Red activities and then create and evaluate employment options for these new CAEN elements. The subordinate commands may also develop considerations for specific targets, as well as subsets of target categories. Also simultaneously or sequentially, the reachback centers can be analyzing different elements of assigned portions of the CAEN to further refine employment options and evaluate their success, ROE violations, and additional target considerations for both specific targets and subcategories of targets. This ability to process in parallel helps reduce the Blue OODA loop, thereby gaining an advantage over the opposing side's OODA cycle.

One useful component of this IO planning methodology is to have a Blue Capabilities Matrix appropriate to each echelon of planning. As shown in Figure 5.2, the national or theater-level command may have a DIME-by-PMESII BCM, which shows how the instruments of national power can potentially be used against each element of the targeted nation's capabilities. At the Joint Task Force command level, the BCM for IO may appear as shown in Figure 5.1, at a level of detail appropriate for considering options available at the JTF level. As one proceeds down the command echelon to specific IO weapon systems or the reachback centers, the level of detail in terms of depth of the BCM increases, while the breadth of options to consider decreases. For example, a specific weapon system may have its own BCM to describe the Red activities against which it can perform desired effect, and where the options might include different types of employment or conditions under which the weapon system might be applied. This scalability ensures that the mapping of the next lower BCM is well defined to the next higher BCM, so that there is no confusion as to where the lower echelon's capabilities fit into the overall option space. Moreover, one can use the tree of options, looking down through lower echelon BCMs, as an increasingly refined set of options for planners to consider. This also helps IO acquisition planners determine required capabilities and the justification for those capabilities.

The reachback centers can be included in the IO planning process through a request for analysis (RFA) process. An RFA is similar to an RFI, which in the intelligence community stands for a request for information. The RFI evolved to address the problem of how military planners could obtain information on a given potential target from the intelligence community when they had no direct authority to task intelligence assets. In a similar manner, the planner contacts the reachback center for help in performing detailed analysis of a particular topic using an RFA.

For example, a planner may desire to create a desired effect on the electric power distribution capabilities of a targeted nation. The planner may not have the expertise to understand all of the intricacies of power distribution, including voltage and phase management. So the planner sends an RFA to a reachback center located in the United States, such as the Joint Intelligence and Planning Capability (JIAPC). An analyst who is an expert in power distribution systems performs an analysis of the targeted nation's electric power system, determines the most effective way to achieve the desired effects without the undesired effects, and sends that information back to the planner.

While both the planner and the analyst could perform this information exchange using a collaborative tool such as the Info WorkSpace (IWS), there is no guarantee that the request, response, and rationale for the response will be stored with the plan. If the planner just puts the answer into the plan or COA, there is no audit trail as to what this answer was based on or its associated assumptions. Moreover, there is no guarantee that the analyst will address all of the planner's concerns in a single pass without some standardized checklist of the factors that need to be addressed, such as rules of engagement or target considerations. The RFA process was designed to ensure that the information for the request provides sufficient detail for the analyst to perform the analysis, and that the response provides sufficient detail for the planner to insert the estimated outcomes into the plan. These estimated outcomes include the probability of success, secondary effects, ROE violations, and other target considerations, as well as relevant point of contact information so that further clarification can be readily obtained. Moreover, all of the rationale associated with the request and response is stored directly with the COA or plan element to which it applies. This makes COA evaluation and plan review much easier, as the necessary supporting data is readily accessible directly from the plan.

Overall, the offensive IO methodology is scalable to all theater-level and operational-level echelon IO planning cells, including the reachback centers. It addresses both higher-level (conceptual) and employment option COA analysis and supports efficient exchange of information between all collaborative planning cells, including those that are remotely distributed. It also addresses the fact that the planning staff may not have all of the information or expertise required to fully generate all of the required elements of the IO plan and ensures that the supporting commands and reachback cells have an efficient and auditable method for providing and tracking the analysis and information provided to the planner.

5.7 Multilevel Security and Distributed Plans

There are two primary reasons why IO planning requires the ability to handle multiple levels of security. The first is that the United States normally fights as part of a coalition of forces, so we need to be able to share some of the planning information with our allies. However, allies in one conflict may be enemies in the next, so how much we choose to share with an ally is a function of who that ally is and our ongoing relationship with them. For example, while Syria was part of the Coalition against Saddam Hussein during Operation Desert Storm, it was trusted with very little in the way of operational plans due to our past history with that nation. Conversely, much information was shared with the our NATO allies in Operation Desert Storm, and at least one member of the Saudi Arabian armed forces was given access to highly sensitive U.S. command, control, and intelligence capabilities [8].

The second reason why IO planning requires multiple levels of security is that many IO capabilities are very fragile. That is, if an opponent knows about a certain IO capability, it can take steps to mitigate that capability, or worse, manipulate that capability against us. For example, during the Kosovo campaign, the Serbian forces were very good at manipulating NATO's electronic eavesdropping capabilities to cause NATO to bomb empty or dummy sites because they learned about the U.S. capabilities used against the Iraqis during Operation Desert Storm [11]. Many IO capabilities require a lot of time and money to develop, and if an opponent were to find out about a capability, that opponent could quickly negate our investment in that capability, as well as the benefits that capability could provide to accomplishing the mission. That is why many IO capabilities are protected by compartmented security procedures to ensure that only the minimum number of people necessary know of its existence.[5]

A number of technologies that provide multilevel security are currently in operation in several specific locations and applications. For example, Oracle version 8i and later provides row-level security so that users only see data appropriate to their level of clearance. The Trusted Network Environment [12], the Information Support Server Environment Guard (ISSE GUARD) system [13], providers of persistence layers such as Hibernate [14] (which explicitly define the rules of who sees what), and modern portal technology can all provide similar capabilities that only show users the level of information appropriate to their level of clearance. These capabilities are being used in certain locations for certain limited applications, but no single application of these solutions has been accepted for broader use.

---

5. That number always includes members of the executive and legislative branches of the U.S. government so they can maintain proper oversight over those capabilities.

The reason for the lack of widespread acceptance of existing multilevel security systems is that the security community often refuses to accept existing technological solutions. The problem is a security policy that currently requires human review prior to releasing any information, which precludes any sort of automated or machine-to-machine interaction using multilevel security systems. There are some valid concerns about whether the technology will always work in every situation, and so far the security community has not been satisfied with any of the technical solutions in more than a few isolated cases.

Until this problem of automated management of multiple levels of security is solved, some sort of work-around will be required to ensure that IO planning involving capabilities that are inherently vulnerable to compromise can be performed in a coalition environment. To this end, we propose the following approach for handling multiple levels of security during IO planning.

First, all of the information at higher levels of classification should remain at the location where that level of classification can be handled. In the case of an IO weapon system, the information about the details of how it works would never leave that location. What is passed to the higher planning echelon are the expected results, included the resulting desired effect, the probability of success, the degree to which the desired effect is likely to be achieved, any ROE violations, and any specific target considerations. With only this generic level of information being passed to the next higher planning cell, the features that make the IO weapon fragile are sufficiently protected.

Second, as the next higher planning cell passes the expected results of their planning to the next higher planning cell, even these expected results are further "rolled up" into a more aggregate set of actions intended to achieve that part of the plan. By the time the accepted COA or plan reaches the level of the Coalition headquarters, all that planners at this echelon will see are the higher-level cause-and-effect relationships of aggregated plan elements, and no details whatsoever of the IO weapon system providing its support to one or more of the sub-subordinate plan elements. At the same time, since the worst-case ROE violations and target considerations are always passed up the chain, no bad news is hidden in the aggregation process. While the aggregation process protects the fragile capabilities, the planner and plan reviewers have sufficient information to understand whether a rule of engagement has been violated and whether any target considerations have been ignored for any target.

Figure 5.12 provides a conceptual schematic of how the data associated with each echelon of planning can be maintained at the appropriate level of detail, so that only the final expected outcomes appropriate to the highest level of planning (the multinational headquarters of the Coalition) can be shared at the secret collateral level.

If the whole plan were developed at the same level of classification, then there would be no need to keep different parts of the plan at different levels of

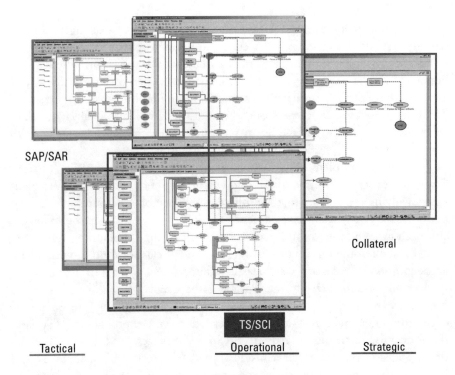

Figure 5.12 Sample display of COA elements at different echelons and classification.

classification. However, since there is a need to keep different parts of the plan at different levels of classification, then, in effect, there is no one place where the whole plan can actually be stored. Since it is important to maintain a complete audit trail of all planning considerations, a way is needed to be able to locate all of the parts of the plan for review by a person who has the appropriate levels of clearances.

The first step to solving this audit trail problem is to recognize that when parts of the plan are physically stored at different locations, we are dealing with a *distributed plan*. It is not just the planning that is performed at distributed locations around the globe; the actual *end product* is a plan whose elements are stored around the globe. A plan reviewer with appropriate clearances to review the entire plan must therefore have a way to locate all the necessary components of the plan so that a complete audit trail is maintained.

The second step is to define a set of plan connectors that let the owner of the higher-level plan know where the supporting elements of the plan are located and who the point of contact is (by person or organization). These plan connectors provide sufficient information as to where one can go (through appropriate channels) to access the more detailed plan elements that are stored at higher levels of classification.

Overall, IO planning requires multilevel security capabilities to perform coordinated planning at all echelons of planning. While multilevel security technology exists, it is not fully exploited to support machine-to-machine information exchanges. To support MLS IO planning, a workaround must be employed that allows different organizations to perform their part of COA and plan development at higher levels of classification, and still provide the necessary plan elements and planning factors to the higher-echelon planning cells. The recommended approach is to provide only the expected outcomes in achieving the desired effect without stating how that effect will be accomplished or what will be used, so that the more classified capabilities will be protected. The result is a distributed plan that contains all of the elements of the plan at all levels of classification. The review of the distributed plan is facilitated by a set of pointers that allow a reviewer with the appropriate clearances know where to find the rest of the plan elements.

References

[1] Joint Publication 3-13, *Information Operations*, February 13, 2006.

[2] Joint Forces Staff College, *Joint Information Operations Planning Handbook*, July 2003.

[3] Ayres, Thomas E., "'Six Floors' of Detainee Operations in the Post-9/11 World," *Parameters*, Vol. XXXV, No. 3, Autumn 2005.

[4] "Operation Praying Mantis," *Wikipedia*, http://en.wikipedia.org/wiki/Operation_Praying_Mantis, February 2006.

[5] Turner, Brian, "After the Civil War, the United States Sued Britain for the Sea Raider *Alabama*'s Depredations," *Military History*, April 2003.

[6] Corley, Lynn, "Correction of Deficiency Reports," Results of Developmental Test November 2004, letter dated November 29, 2004.

[7] Kent, Glenn A., and William E. Simmons, *Concepts of Operations: A More Coherent Framework for Defense Planning*, RAND, N-2026-AF, 1983.

[8] Schwarzkopf, General H. Norman, *It Doesn't Take a Hero*, New York: Bantam Books, 1998.

[9] Carhart, Tom, *Iron Soldiers*, New York: Pocket Books, 1994.

[10] *Glossary*, Naval Warfare Development Center, http://www.nwdc.navy.mil/Library/Documents/NDPs/ndp5/ndp50006.htm.

[11] Thomas, Timothy, L., "Kosovo and the Current Myth of Information Superiority," *Parameters*, Spring 2000.

[12] *The Trusted Network Environment*, General Dynamics Advanced Information Systems' public Web site, http://www.gd-ais.com/Capabilities/productpages/tne.htm, December 2005.

[13] *The Information Support Server Environment GUARD*, ISSE Guard public Web site, Air Force Research Laboratory, http://www.rl.af.mil/tech/programs/isse/guard.html, December 2005.

[14] "Relational Persistence for Idiomatic Java," Hibernate homepage, http://www.hibernate.org/, February 2006.

6

Planning Methodologies and Technologies for Influence Missions in IO

In war, Truth is the first casualty.
—Aeschylus, 525 B.C.–456 B.C.

Of the five core capabilities of IO described by Joint Pub 3-13, psychological operations (PSYOP) is the core capability for influence operations, while public affairs, civil military operations, and defense support to public diplomacy are related capabilities. (Note that the U.S. Air Force defines influence operations differently, which is discussed in Section 6.1.) According to JP 3-13 [1], "PSYOP are planned operations to convey selected information and indicators to foreign audiences to influence their emotions, motives, objective reasoning, and ultimately the behavior of foreign governments, organizations, groups, and individuals. The purpose of PSYOP is to induce or reinforce attitudes and behavior favorable to the originator's objectives."[1]

The commander wants to understand how influence operations can help accomplish the mission. The planner needs to be able to provide the commander with the key elements of what an influence operation is trying to accomplish and how it fits into the overall plan. The planner also needs to explain how

---

1. JP 3-53 uses the term *indicators* to refer either to part of a message, or to results or responses that indicate whether your message is being accepted (e.g., indicating resolve as a response to attention-getting). The methodology in this book distinguishes between these two definitions [1].

nonkinetic influence operations can leverage kinetic actions to accomplish influence objectives and how influence operations can help accomplish the objectives of kinetic actions.

This chapter is organized as follows. Section 6.1 discusses the differences between Joint and Air Force doctrine relating to influence operations and explains the categorization of different IO components with respect to whether they are more closely aligned with influence or defensive capabilities. Section 6.2 categorizes the *purposes* of messages, how these purposes help guide the influence operations planning process, and how influence planning fits into the overall planning process. Section 6.3 introduces the *idea battlespace*, which is a conceptual arena in which messages compete for dominance. The idea battlespace is the equivalent of the battlefield for influence operations and defines the *context* in which all influence operations are undertaken. Section 6.4 describes the different *methods* by which messages compete for *attention* and try to maintain that attention over time. Without attention, messages are not perceived (or accepted) by the intended audience. Section 6.5 presents the different *change methods* that an influence operations planner can attempt to apply to the state of an enemy or friendly message within the arena. This section defines the list of ways in which one message can gain leverage over another in order to gain *dominance* with the intended audience.

Section 6.6 discusses various *delivery mechanism* options by which messages are physically delivered to the intended audience. This section also notes the importance of shifting some of the emphasis of doctrine development and funding away from simply improving delivery mechanisms and toward improving planning and capabilities to better compete within the idea battlespace.

Section 6.7 discusses the three different *measures of success* associated with influence operations that are essential to IO planning. In particular, the *measure of causal linkage* is essential to the success—and the measurement of success—of influence operations.

Section 6.8 shows how all of the preceding material can be entered into the cause-and-effect network (CAEN) described in Chapter 5. This allows influence operations to be planned and compared on an equal footing with IO attack options, as well as with more traditional kinetic options. Thus, influence operations, offensive IO, and traditional kinetic capabilities can all be included in the same plans using the same planning process and comparison methods to select the best course of action to accomplish the mission.

6.1 The Influence Operations Planning Context

According to Joint doctrine and the DoD IO Roadmap, influence operations are primarily psychological operations and are distinct from military deception and operations security. Both Joint Publication 3-13 and DoDD 3600.2, "Information Operation Policy," define the five core IO capabilities as electronic warfare (EW), computer network operations (CNO), psychological operations (PSYOP), military deception, and operations security (OPSEC) and describes public affairs (PA) and civil affairs as IO-related activities. In contrast, the most recent version of Air Force DD 2-5 divides the core IO capabilities into electronic combat operations (including traditional EW), network combat operations, and influence operations. Within influence operations, the Air Force currently includes PSYOP, military deception, OPSEC, public affairs (PA), counterpropaganda, and counterintelligence [2].

The primary reason for this difference is that the Air Force is putting most of its efforts into the two areas where they have traditionally had the most strength and interest (electronic combat operations and network combat operations) and lumping everything else into the "other" bin of influence operations. From the Air Force perspective, this makes organizational sense. However, it does not agree with Joint doctrine, and, in the opinion of this author, also creates some undesirable side effects.

One can argue that military deception and operations security are influence operations because they attempt to influence the enemy to not see what the friendly side is trying to accomplish. However, military deception and OPSEC are primarily passive or defensive efforts, with the primary objective of preventing the enemy from interfering with your operation. Most other influence operations, including PSYOP, public affairs, and counterpropaganda, are all active approaches designed to change the behaviors of the targeted audience. Therefore, from a planning process perspective, PSYOP, public affairs, and counterpropaganda are primarily influence operations. Military deception and OPSEC planning are more similar to defensive IO planning, as described in Chapter 7.

Note that public affairs differs from PSYOP for two reasons. First, public affairs is not supposed to tell an intentional lie, but is restricted to telling the truth as best as it is known at any given time. PSYOP, military deception, and OPSEC are under no such restrictions, and lying about one's location to the enemy is perfectly acceptable behavior. That is one reason why U.S. doctrine for psychological operations precludes PSYOP from being intentionally employed against the U.S. public, because PYSOP is not legally precluded from lying.

The latest Air Force IO doctrine will eventually need to be reconciled with the Joint IO doctrine. Even though Joint and Air Force doctrine divide the components of IO into different bins, this book provides planning techniques for each of those components. Since this book places MILDEC and OPSEC planning as separate from PYSOP, it therefore follows the IO roadmap and JP 3-13 more closely than does Air Force doctrine at this time.

The rest of this chapter presents a fairly new approach to influence operations planning, and focuses primarily on PSYOP and public affairs planning. To a great degree, this new methodology places more emphasis on previously obscure parts of PSYOP (such as gaining attention and clearly defining each message's purpose) and also adds new concepts such as the idea battlespace to better visualize the conflict space. In particular, we define a set of categorizations that are useful to the PSYOP and public affairs planner. For example, defining the four categories of PSYOP and public affairs message purposes helps the planner and decision-maker better understand the purpose (desired effect) of a given message and how to measure the achievement of that desired effect. In a similar manner, categorizing the various change methods available for friendly and enemy messages helps keep the scope of options manageable for the PSYOP and public affairs planner. It also helps the public affairs planner understand which change methods are always forbidden or in which ways some change methods are forbidden. This clarity of message purpose and change methods available also assist defense support of public diplomacy.

6.2 Categorizing Message Purposes

Based on discussions with field-grade officers, general officers, and members of the PSYOP community, it appears that the PSYOP community is having some difficulty explaining what PSYOP can do for the commander and the accomplishment of his mission. To this end, in this section we offer several key definitions that will help provide a common basis for discussion so that planners can better integrate nonkinetic influence operations with the more traditional kinetic elements of a plan.

First, we define some terms that will be used throughout this chapter. These definitions provide the context of the actors (groups and messages) within the idea battlespace, the interactions between these actors, and a description of the idea battlespace itself.

- A *group* is a well-defined population set of interest to planners and relevant to the objectives of the plan. The term used for *group* in JP 3-53 is target audience (TA); however, not all communities like the terms *target* or *audience*. We use the term *group* to define the intended recipient of the message; this term can readily be changed to whatever the PSYOP community ultimately decides to use.

- A *message* is an idea expressed with the intent or purpose of influencing beliefs and behaviors in one or more groups.

- An *action* is taken to deliver the message and/or to gain/maintain attention. For example, the combination of a change method and change option defines an action. JP 3-53 also categorizes the employment of a delivery mechanism as an action; the methodology in this book distinguishes between the two uses of the term.

- An *interaction* can be between messages and groups, messages and messages, and groups and groups.
- The *idea battlespace* is a conceptual arena, formed by the groups of interest, in which ideas compete with one another for dominance. This concept will be described in detail in Section 6.3.

In the idea battlespace, everything interacts with everything else, making this a relationship-intensive environment. Moreover, an infinite number of messages can be created, which further expands the space of options. To help mitigate the size of the problem, we have defined a set of categorizations for PSYOP that enable planners to determine the purpose, intent, or objective of a message. Further, and as an additional departure from current doctrine, we explicitly distinguish between the *purpose* of the message and the message itself. An influence message may have one of four purposes:

- Cause Group A to Believe X;
- Cause Group A to Not Believe X;
- Cause Group A to Take Action Y;
- Cause Group A to Not Take Action Y.

Depending on the circumstances, it is sometimes easier to convince someone to not believe something than to believe something, and it is sometimes easier to get a person to not do something than to do something, especially when the action taker assumes increased risk.

This simple categorization of message purpose helps to keep the planner and reviewer focused on the purpose of the message, so that the subsequent crafting of the message and selection of delivery mechanism can then be compared to the purpose of the message for consistency. It also helps categorize the purposes of different messages in the idea battlespace for ease of comparison. For example, a user's filtering and focusing on all messages that support or counter a particular message purpose are quick ways to select messages relevant to a given mission.

Note that a given message may have multiple purposes. For example, a single terrorist videotape broadcast may include the following purposes:

- Cause Americans to believe that they are vulnerable;
- Cause Islamists to believe that al-Qaeda is good;

- Cause non-Americans to disbelieve anything the United States says;
- Cause young fanatics-to-be to take action against the United States;
- Cause U.S. allies to not take action against al-Qaeda agents for fear of reprisal.

Of course, the original message can be divided into a set of submessages, where each submessage has one and only one purpose.

This categorization of the four purposes or intents for messages helps keep the option space for the influence operations planner more manageable and easier to explain. Moreover, it will help the planner more clearly define the idea battlespace, described next.

6.3 The Idea Battlespace

The *idea battlespace* is a concept whereby groups of interest form an arena in which ideas (messages) compete with each other for dominance. Groups are added or removed from the arena over time as messages or group interactions bring groups in or move them out.

Within the Idea Battlespace arena, messages fight against other messages for dominance, much like programs compete against each other in the movies *Tron* or *The Matrix*. Message dominance is defined by the level of attention and the level of belief achieved by the message with respect to the groups of interest that define the arena.

The level of belief or disbelief in a message, relative to other messages, shifts over time. As messages interact with one another, a message that is dominant at one time may not be dominant at another. A major factor in this changing relationship among messages is shifts in the *level of attention* a message receives from a target group or groups. An important key to success in the idea battlespace is bringing attention to your message and maintaining that attention.

The idea battlespace is a conceptual tool designed to help planners focus on the groups and messages relevant to the operation at hand, while maintaining the global context of the potential impact of messages. Just as many messages may be competing within a single idea battlespace, many separate idea battlespaces may be active at a given time. Determining where to place the boundary with respect to a given idea battlespace is a function of the scope and responsibilities in the various planning cells.

Figure 6.1 shows a sample idea battlespace with groups A through F. (Group G enters the battlespace later.) These six groups of interest are viewing the friendly messages (FR-###) and enemy messages (EN-###) in the arena.

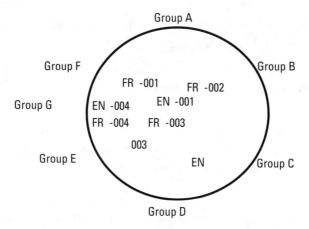

Figure 6.1 Sample idea battlespace display.

Each group will have a level of belief or disbelief in each message in the arena. Moreover, how each group perceives other groups will also influence its acceptance or rejection of each message from that group.

In addition, one side can create a message (EN-004) that not only plays in this arena, but may drag another group (G) into the arena as well. The other side(s) may then need to adapt their messages or create new ones to respond to this new participant in the arena (FR-004). For example, when al-Qaeda declared that the United States was waging a war against all of Islam, that message brought all Islamic nations and Muslim groups into the idea battlespace. The U.S. message reassuring nonradical Muslims that the fight was only against Islamic terrorists helped keep the conflict from escalating, which had been the intention of the al-Qaeda message.

The current state of an idea battlespace can be described by defining the following attributes:

- The groups of interest that form the current idea battlespace arena;
- The messages and purposes of messages that are competing within the arena;
- The interactions of messages with groups;
- The interactions of messages with messages;
- The interaction of groups with groups;
- The current level of attention on each message as viewed by each group.

Defining the groups that make up the arena has been described earlier. Defining the purposes of each message was described in Section 6.2. The level of attention given to messages is described in more detail in Section 6.4. The remainder of these six features is described next.

6.3.1 Messages Interacting with Groups

Note that influence operations planners currently perform the research and analysis step as defined in JP 3-53 to prepare their situation awareness [3]. The research and analysis step includes determining "competing and complementary US and other-party goals, and possible strategies and PSYOP supportability of COAs to achieve these goals" [3]. This step also includes gathering information on ethnic preferences and biases, symbols, and other important factors in message development and presentation.

The idea battlespace is a newer construct that expands on the component steps of JP 3-35 in order to more clearly define the key relationships between messages and groups. Each message has one or more clearly stated purpose, each group has a view of each message, messages interact with each other, and groups interact with each other. All of these relationships need to be defined to some level of detail to adequately understand the current idea battlespace situation. This level of detail used to describe the status of these interactions may be limited simply to "Group A's level of belief in Message FR-001 is high," but even this provides a means of documenting that estimate for others to review and base their plans upon.

Table 6.1 shows how one message (FR-001) is viewed by the various groups in the idea battlespace. The first parameter (first column) is the percentage of that group that believes the message, or the best estimate available to the influence operations planner.[2] The second parameter is the percentage that does not believe it. The third parameter is the percentage that does not know or does not care. The fourth parameter is the level of attention the message has within the entire set of messages viewed by that group (i.e., the level of attention this message is receiving relative to other messages). The fifth parameter is the level of confidence you have in these estimates (similar to the intelligence ranking of confidence). The second-to-last column is the date at which the estimate of these parameters was made and the last column is the source of the estimate.

As an alternative to estimating the level of acceptance of one message as viewed by all groups, one can also estimate the level of acceptance of *all* messages as viewed by *one* group. Table 6.2 shows a different slice of the idea battlespace,

---

2. Given the fuzzy nature of measuring belief without a scientific sample, this approach does not require planners to know precise numerical percentages; it only requires that planners enter some text-based or numerical estimate into the cell. If the best estimate of the level of belief is high, then the planner would enter that word into the cell. If a scientific sample is available, then the numerical value should be entered.

Table 6.1
Level of Acceptance of a Message (Fr-001) Across Different Groups

Group ID	Percentage That Believes	Percentage That Does Not Believe	Percent That Does Not Care	Attention Ranking	Confidence	Date of Estimate	Source of Estimate
Group 1	85%	5%	10%	High	High	01/01/03	XXX
Group 2	20%	5%	75%	Low	Medium	03/01/03	YYY
Group 3	10%	60%	30%	Medium	Medium	05/05/03	ZZZ
Group 4	0%	90%	10%	High	Low	06/10/03	WWW

Note: If the purpose of the message is to have each group take an action, then replace "Percentage That Believes" with "Percentage That Are Likely to Take Action," and so forth.

Table 6.2
Level of Acceptance of a Set of Messages as Viewed by a Single Group

Message ID	Percentage That Believes	Percentage That Does Not Believe	Percentage That Does Not Care	Attention Ranking	Confidence	Date of Estimate	Source of Estimate
Fr-001	85%	5%	10%	3	High	01/01/03	XXX
Fr-002	70%	10%	20%	2	Medium	04/11/03	AAA
En-001	20%	60%	20%	1	Medium	05/15/03	BBB
En-002	10%	30%	60%	4	Low	06/08/03	CCC

Note: If the purpose of the message is to have each group take an action, then replace "Percentage That Believes" with "Percentage Likely to Take Action," and so forth.

focusing on how one group views all of the messages currently in the idea battlespace. The parameters for this table are the same as in the previous display.

One could also define and display all messages as viewed by all groups across a set of parameters. Table 6.3 displays the first four parameters (Percentage That Believes, Percentage That Does Not Believe, Percentage That Does Not Care, and Level of Attention) in four subcells of the matrix of messages (rows) viewed by groups (columns). The user should be able to select the elements that make up the legend, and all parameter elements selected can be displayed in this format.

6.3.2 Messages Interacting with Messages

Messages that neither conflict nor support each other are considered *orthogonal* to each other, as if they were neutrals in a conflict between two other messages.

Table 6.3
Display Type 3: Level of Acceptance of a Set of Messages Among Selected Groups

Group ID by Message ID	Group 1				Group 2				Group 3				Group 4			
Fr-001	85%	5%	10%	3	20%	5%	75%	Low	10%	60%	35%	3	0%	90%	10%	2
Fr-002	70%	10%	20%	2	25%	15%	60%	Low	30%	65%	5%	1	20%	75%	5%	3
En-001	20%	60%	20%	1	75%	5%	20%	Medium	70%	5%	25%	4	70%	5%	25%	4
En-002	10%	30%	0%	4	60%	10%	30%	Medium	65%	5%	30%	2	50%	40%	10%	1

Note: If the objective is to take an action, then replace "Percentage That Believes" with "Percentage Likely to Take Action," and so forth.

% Believe	% Don't care	% Not believe	Level of attention

Conflicting messages are ones that contradict each other in some way. Planners need to look for conflicts and orthogonality among all messages and document these attributes for review and planning purposes. They also need to look for consistency and completeness in friendly message sets by identifying conflicts, synergy, and gaps. In a similar manner, the planner needs to look for inconsistency and incompleteness in enemy message sets by identifying conflicts, synergy, and gaps that can be exploited.

Messages issued by many opponents of the United States are often wildly contradictory. This often does not matter to audiences unfriendly to the United States, who are often willing to believe any anti-U.S. message even if it directly contradicts another such message. The rantings of "Baghdad Bob" were examples of self-contradictory messages that were believed in the Middle East, but considered humorous in the West [4]. The United States could still exploit these contradictions for Western audiences even though the Middle Eastern audiences did not care that they were contradictory.

6.3.3 Groups Interacting with Groups

Group-to-group interactions are important measures in the idea battlespace. These interactions depend upon, and often define, cultural, ethnic, and historical biases and predispositions and the relative influence of one group over another. Group-to-group interactions can greatly affect the acceptance of a message. For example, who delivers the message matters in whether and how it is accepted. Thus, if Group A has a long history of ethnic conflict with Group B, Group A will be unlikely to believe Message X if it is delivered *by* Group B. Similarly, if Group A believes message X, Group B will *not* believe message X simply because Group B hates Group A. Table 6.4 is a sample display of group-to-group interactions. Graphical types of displays are also useful, in addition to tabular displays. Group-to-group interactions are often better represented by an influence diagram or a directed graph than a table.

Social network analysis (SNA) tools can help in identifying and quantifying these relationships. A sample social network analysis tool was described in [5].

6.4 Getting and Maintaining Attention

The level of attention a group gives to a message is largely a function of the amount of attention *brought* to that message. Activist groups have understood for years the value of attention-getting actions to bring attention to their message. Nations have similar attention-getting methods. As shown in Figure 6.2, the scale of attention-getting escalates from words to protests to damage to destruction to loss of life to war.

The media plays a major part in which messages get attention, how much, and whether that attention is positive or negative. As discussed briefly in Chapter 1, Western media have well-established priorities in reporting the news. Fatalities are always considered more newsworthy than casualties, casualties are more newsworthy than damage, and so on down the line. Thus, with respect to Figure 6.2, Western media reporting typically works from the upper right corner of attention-getting methods to the lower left corner. Radical groups

Table 6.4
Group-to-Group Interaction Table

	Group 1	Group 2	Group 3	Group 4
Group 1	X	Very positive, somewhat subordinate	Very negative	Somewhat negative
Group 2	Very positive, somewhat dominant	X	Somewhat negative	Somewhat negative
Group 3	Very negative	Very negative	X	Very positive, subordinate
Group 4	Very negative	Very negative	Very positive, dominant	X

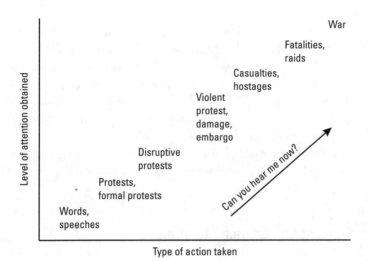

Figure 6.2 Spectrum of attention-getting actions and escalation ladder.

understand these media priorities and have frequently taken escalatory action to gain increased attention to themselves and their messages.

Attention is perishable. Getting attention is fairly easy, but maintaining attention over an extended period of time is difficult. This is particularly true for U.S. audiences, whose attention span is notoriously short. If you do not keep attention on a message, another message will take over the top place for attention in that group. Repeated and varied actions help maintain attention on a message, while escalation of attention-getting methods increases attention. Conversely, too much repeated attention-getting in the same way can desensitize the group to the message (attention fatigue).

Note that gaining attention is the primary way that kinetic options can be used to help achieve the objectives of an influence operation. Kinetic options, such as dropping bombs or the threat of dropping bombs, can be tied to a specific message, such as "Abandon this position and save yourself." According to Al Zdon [6],

> One strategy was to warn the Iraqis of an impending bombing attack and then follow through. "We would tell a particular unit that we would bomb them within 24 hours and they must leave. A lot of them would do just that. We would follow through on the bombing so they knew we were telling the truth."

6.5 Change Methods and Submethods

Just as the term *effects* in IO applies to methods used to change the current state of a target into the desired state, *change methods* are used to implement changes within the idea battlespace. Change methods include the mechanisms or approaches by which any member in the arena of the idea battlespace can create a change to that arena. There are currently eight types of change methods that can be applied to messages within the idea battlespace; each includes a set of options, or submethods, that can further refine the mechanism for creating change. The eight change methods are to:

- Create a new message;
- Modify an existing message;
- Change the level of attention on a message;
- Co-opt an existing message;
- Subvert a message;
- Distract attention from a message;
- Ignore or provide no response to a method (silence);
- Restrict access to a message.

The first three change methods involve actions taken on friendly messages, either by creating a new message, modifying an existing message, or changing the level of attention on an existing message. The last five change methods involve actions taken against enemy messages. Note that of these five methods, the first three require the creation of a new friendly message to achieve the desired effect.

6.5.1 Creating a New Message

Creating a new message is the traditional way to affect change in the idea battlespace and is the focus of most of the current PSYOP doctrine. One particular benefit of creating a new message is that if it is the first message in the arena on a given subject, it tends to set course for the debate that follows. As in most military combat situations, initiative is important in the idea battlespace. Being first with a message places your opponents in response mode, requiring them to react to your message.[3] Unfortunately, due to the long preparation and review time currently required for most U.S. messages, we often forfeit this initiative to our opponents.

---

3. An alternative to responding to an enemy message is to ignore it and focus instead on your own message campaign, if you can afford to do so. See the following silence change method.

6.5.2 Modifying an Existing Message

It may be necessary to tailor a message for better acceptance by one or more groups, to change the emphasis on a portion of an existing message, or to improve consistency among messages. Sometimes this approach involves clarifying an earlier statement (by issuing a new message); at other times it may involve changing the message slightly to better compete in the idea battlespace. For example, President Bush's initial major speech involved the word "crusade" in the war against terrorism as a statement of determination. This word has never been repeated in subsequent speeches because the terrorists were twisting the word to imply it is a Western crusade against all of Islam. Of course, proterrorist Web sites continue to replay that initial speech repeatedly to imply the war is a West-versus-Islam conflict in spite of all facts to the contrary [7].

6.5.3 Changing the Level of Attention

Changing the level of attention upon a specific message is a way to create change in the idea battlespace. For example, connecting a violent action with a message makes a greater impact than the message without the action, or the action without the message. This is one reason why the anti-WTO demonstrations are intentionally violent and destructive—to gain more attention than would a nonviolent demonstration [8]. The use of violence to gain attention has also been called "propaganda of the deed" in anarchist and other antiestablishment literature [9]. Section 6.4 described various ways of generating or maintaining attention upon a message.

6.5.4 Co-Opting an Existing Message

This is the first of the five change methods that involve actions against enemy messages. Note that this change method, as well as the next two, requires the creation of a new friendly message to achieve the desired effect.

There are two ways of co-opting an existing enemy message. One is to genuinely adopt the message as our own, such as the United States becoming a champion of human rights. When President Jimmy Carter adopted support for human rights as stated official policy, it not only carried a lot of weight for the cause of human rights, it was a genuine effort on the part of the United States to support human rights around the globe. Note that the United States had previously supported human rights, but had not made it an official part of our foreign policy.

The second way to co-opt a message is to claim ownership by pretending to take ownership of the message even if one is not actually doing so. For example, a number of U.S. Muslims donated to Islamic charities that claimed they were going to use the funds for humanitarian aid to Muslim refugees. In reality, a number of these Islamic charity organizations were fronts or unsuspecting contributors to terrorist organizations, which siphoned funding from unsuspecting donors [10]. These false charities co-opted the message of humanitarian assistance into funding acts of terror.

6.5.5 Subverting a Message

There are many ways to subvert a message, including discrediting the messenger, sowing doubt, focusing attention on an extreme part of a message, associating the message with a fringe element, lampooning it, or just plain lying about it. Attacking and discrediting the messenger are one of the most common ways of subverting a message. The U.S. tobacco industry was a master of this technique. For example, for several decades, any expert who demonstrated the risks of smoking was accused of misconduct or of being a charlatan [11]. An inside whistleblower who released thousands of internal documents describing industry collusion was accused of being a petty thief and of attempting to extort money from the industry. Expert witnesses were accused of either being mercenaries or having an agenda to destroy free enterprise [12].

Another way to subvert a message is to sow doubt and confusion about the message itself. According to historian Robert Leckie, Hitler used his "Fifth Column" to sow the seeds of demoralization and doubt about the rightness or the ability of a nation to defend itself against the Nazi advances [13]. Moreover, placing doubt on a message, no matter how well supported by scientific or factual evidence, will encourage target groups to believe what they prefer to believe. As in the previous example, the tobacco industry was also a master of sowing doubt about the scientific evidence that smoking causes cancer and is addictive. For decades, they trumped up scientific evidence supposedly countering actual scientific evidence using labs that would produce any result for the right price. To quote Corporate Accountability International [14, 15]:

> For nearly 40 years, the industry-sponsored Council for Tobacco Research (CTR) has waged what the *Wall Street Journal* recently labeled "the longest-running misinformation campaign in US business history." Portrayed as an independent scientific agency to examine "all phases of tobacco use and health," the CTR has actually been the centerpiece of a massive industry effort to cast doubt on the links between smoking and disease. According to Michael Pertschuk, former chair of the Federal Trade Commission, "there has never been a health hazard so perfectly proven as smoking, and it is a measure of the Council's success that it is able to create the illusion of controversy in what is so elegantly a closed scientific case."

Another way to subvert a message is to focus on only one part of the message that may be less appealing to portions of the targeted groups. If the message has a single feature that could be emphasized (usually a negative one), that feature can be used to outweigh any good features of a message. For example, if a social organization is trying to start a halfway house in a middle-class neighborhood, some opponents have used the argument that such a place would only attract "undesirable elements," thus trying to convince the community into not allowing the facility to be built [16].

Another, more complicated method, of subverting a message is to find a fringe element that supports a given message and then give that group lots of coverage. By focusing on fringe groups that (falsely) claim to be speaking for the larger group, the media can shift popular support away from an otherwise valid message through guilt by association. For example, Muslim groups (correctly) pointed out that the actions of radical Islamic terrorists were causing the entire Islamic faith, and its adherents, to be associated with the violent actions of a small fringe group [17]. This is the converse method to discrediting the messenger, as it uses the association of an already discredited messenger to discredit a valid message.

Yet another way to discredit a message or messenger is to lampoon it. Political cartoons have a long and effective track record of guiding popular opinion. The tobacco industry frequently hired or otherwise incentivized cartoonists, columnists and actors to lampoon antitobacco efforts or litigation efforts in general [18]. Making a joke of a message or messenger is a very effective tool against a message—even if the lampooning is unfounded.

Lying about a message is probably the most common and effective way to subvert a message. Although U.S. military public affairs offices are not permitted to use this technique, it is a technique frequently used by our enemies against U.S. messages. Spokesmen such as Goebbels, Baghdad Bob, and Al Jazeera are classic examples of this approach. Unfortunately, this has often been successful in achieving the desired effect upon the target audiences [19]. Goebbels lied about almost everything during World War II, and especially about Allied military successes to the German public to keep up morale on the German home front [20]. In 2003, Baghdad Bob performed a similar function for a similar dictator [21]. In 1991, the Iraqis also lied about Coalition claims to be avoiding damage to cultural sites by fabricating damage to a mosque to claim the Coalition damaged it [22]. Al Jazeera was recently caught using file footage as though it were new "news" regarding a bombing in Dahab, Egypt, and claimed additional bombings in Belbis, Egypt, that never occurred [23]. Since the Egyptian message is that Egypt is safe for tourism, Al Jazeera lied about attacks against tourists that did not actually occur and used file footage from a 2004 bombing to convince people that it did occur.

One form of lying is the unproven accusation (or allegation). An accusation can carry a great deal of weight with a group that is predisposed to believe it, simply by being made and having attention brought to it, even if it is known to be false. For example, the claim that the slow response to the effects of Hurricane Katrina was due to intentional malice toward the black community of New Orleans (rather than simple governmental incompetence) was untrue, but based on subsequent polling results in the black community, it proved to be a very effective antigovernment message [24]. (We discuss the power of the accusation further in Chapter 11.)

Discrediting a message by lying about it is very effective when targeting a group that is predisposed to believe anything negative about the source of the original message, whether true or false. If a group has been indoctrinated to hate the United States, then any piece of negative information about the United States, including outright lies, will generally be believed, and any message to the contrary will be ignored. This is why lies have so much power in the idea battlespace. It is extremely difficult to defend against this means of discrediting a message, for even when one can definitively prove that the opponent is lying, this "fact" may be of no importance to the group targeted by the lie.

Distracting attention from a message is a common technique to reduce the amount of attention on, or influence of, a message that is currently winning out against your other messages. The method is simply to place a different (and hopefully more attention-getting) message on the minds of the populace. Flag-waving and war are common methods to unite the people against a common enemy and distract them from the failings of the current leadership. For example, to distract the population away from its domestic troubles, Argentina decided to take the Falklands (or Malvinas) back from the United Kingdom [25]. The 1997 movie *Wag the Dog* takes a satirical look at this well-known and oft-used principle.

Ignoring or not responding to a message (silence) is another way to compete against a message. Some messages are simply too far out there to justify a response. For example, conspiracy theories that the United States is hiding alien bodies and spaceships in Area 51 are not worth responding to, as responding to them simply continues to focus attention on the outlandish message. Note that this approach does not require the creation of a new, friendly message.

Lastly, restricting access to a message is a way to minimize the effect of a message. In this day of ubiquitous global communications and transportation, however, this approach is becoming increasingly difficult, except in nations with state-controlled media such as China and North Korea. Otherwise, it is very difficult to sufficiently restrict access to unfavorable messages, and in many cases, attempting to block access to a message often backfires. When Soviet Bloc nations tried to jam the Voice of America, the act of jamming caused some folks to want to find out what their governments did not want them to hear [26]. When the United States bombed the Iraqi TV station in Baghdad, the worldwide reaction was very negative [27]. Restricting access to a message in today's globally interconnected world is very difficult, and therefore will likely be less often used in the future.

6.5.6 Constraints on Public Affairs Personnel in the Idea Battlespace

While the idea battlespace and all of the preceding material can also help the U.S. military public affairs planner, the significant difference is that the set of change methods for the public affairs planner is much more restrictive than for PSYOP planners. Since U.S. PSYOP doctrine precludes the U.S. military from performing PSYOP against the American public, U.S. military public affairs (PA) personnel are legally bound to tell the truth as best they know it. Thus, the PA planner *cannot* use the following change methods:

- They cannot knowingly tell a lie.
- They cannot co-opt an enemy message by *pretending* to co-opt that message (i.e., if using the co-opting method, the message must be genuinely co-opted).
- They cannot intentionally sow doubt and confusion in the American public.

PA planners *can* use the following change methods as long as they do not employ any lies as part of those methods:

- Create a new message.
- Modify an existing message.
- Change the level of attention on a message.
- Co-opt an enemy message by genuinely co-opting the message.
- Subvert a message so long as only the truth is used in the process. For example, PA planners can use the discredit-the-messenger submethod of subverting a message as long as the information used to discredit the messenger is truthful and factual.
- Distract from a message using another message so long as the new message is true.
- Ignore or choose not to respond to a message (a method that is frequently used).
- Restrict access only to those messages that are under the initial control of the U.S. military and have not been released by proper authorization, such as the Freedom of Information Act.

PA planners, however, must remain aware that our opponents are being under no such constraints, and are using and will continue to use all of these change methods, especially lying. Thus, PA planners must be prepared to counter all of the change methods described earlier.

6.6 Delivery Mechanisms

As mentioned earlier, current PSYOP planning focuses primarily on message development and delivery mechanisms. For example, the recent PSYOP ACTD focuses on improved delivery mechanisms, and Joint Pub 3-53 emphasizes new message creation and delivery mechanism selection. Since this subject is fairly well defined in current doctrine, this book does not attempt to address the tactical aspects of planning for delivery mechanisms or production management (such as printing leaflets or editing video). We will, however, discuss some of the operational and strategic level influence operations planning considerations.

The influence operations planner needs to know the following information:

- Which delivery mechanisms are available?
- Which delivery mechanisms can reach each target group, and when?
- Which delivery mechanisms are preferred from the perspective of message acceptance for that group?
- How much lead time is required for the chosen delivery mechanism (from start to finish)?
- How cost-effective is the chosen delivery mechanism [28]?
- How will the success of the delivery (vice message acceptance) be measured, and who will measure it and when?

Furthermore, the delivery mechanisms for the message must be synchronized and deconflicted with other aspects of the plan. For example, if the preferred delivery mechanism is a television broadcast on day 2, and the power distribution system will be knocked out on day 1, the message is not likely to reach its intended audience.

6.7 Measuring Influence Success

There are three distinct types of measures to determine the degree of success of an influence operation:

- The successful delivery of the message (MOP);

- The achievement of the desired effect in the targeted group (MOE);
- The assumed or observed *causal linkage* between the action taken and the desired effect on the targeted group (MOC).

The degree of success of actions taken, such as message delivery, is measured by measures of performance (MOPs). These are usually quantitative measures that are readily attainable, such as what number of leaflets reached the target area, whether they were delivered on time, whether the scatter pattern was good, and so forth. MOPs describe how well the action was performed, but do not describe whether the desired effect of that action was achieved.

The measure of success of the desired effect is attained through a measure of effectiveness (MOE). The MOE may be a quantitative measure, such as the number of enemy troops surrendering or the number of deserters, or a more qualitative measure, such as unit morale. Unless one has direct access to a representative sample of the target group, MOEs are often subjective and not directly measurable. For example, the real measures of effectiveness may only be obtained after a conflict is over and participants from the opposing side are interviewed. In other cases, on-the-ground surveys of groups of interest are often being undertaken by groups such as the Pew Institute, and these sampling surveys are good, quantitative measures of the MOEs for the achievement of desired effects [29]. Although a little more indirect, surveys of published media can provide some MOE estimates, although the media is not an exact march with the population's beliefs in any country.

The third measure, the measure of causal linkage (MOC), is more difficult to measure. Since the connection between the action and the desired effect is not an exact science, *an assumed or observed causal relationship* is required to connect the success of the action to the achievement of the desired effect. We say assumed because for planning purposes, one may need to make an assumption that an action will contribute to the achievement of a desired effect even when one has no quantitative data that clearly establishes such a cause-and-effect. In other cases, there may be quantitative or at least anecdotal evidence that a specific action will achieve a desired effect. For example, in the case of leaflets dropped on enemy forces to persuade troops to surrender or desert, on the average, a known number of enemy troops will surrender given a certain number of leaflets (subject to the level of training and morale of the target group). However, even this sort of quantitative information is only applicable if the same conditions under which the previously observed outcomes were achieved apply to the new situation as well. This is not always the case. During Operation Desert Storm, initial leaflet drops did not achieve the anticipated number of surrenders. It was only after the ground war began and Iraqi soldiers began surrendering in droves that we learned that one of the reasons for the small number of early responses was the roving Republican Guard death squads [30]. Once this factor became known, it could be accounted for in subsequent estimations of the number of enemy surrenders per leaflet drop.

If the planner is only dealing with the MOP and MOE, and the MOP looks good but the MOE looks bad, where does one look for the solution to the problem? Explicitly defining a measure of the causal linkage helps the planner focus on the weakest link in the planning assumption—that the action will cause the desired effect to occur. This helps avoid the problem of changing the plan's measure of the desired effect when the correct one is actually being measured, or simply repeating the same types of actions hoping for a different result when the causal relationship is not really understood.

As mentioned in Chapter 3, the link from direct actions to indirect desired effects is a key feature of IO planning. In the case of influence planning, understanding and measuring the causal linkage are an important requirement for successful influence operations planning. Due to the longer planning horizons of influence operations compared to most offensive IO desired effects, there is a need to monitor the MOPs, MOEs, and MOCs over time to ensure that they are progressing in expected ways. Since you cannot manage what you cannot measure, tracking the effectiveness of the MOCs, as well as the MOPs and MOEs, is essential to good IO planning and execution monitoring. Moreover, the required feedback loop to continually improve IO planning factors (as described in Chapter 3) is especially important to the planning, monitoring, and success of influence operations.

6.8 Entering Influence Operations in the CAEN

The planner needs to ensure that the influence operations are fully integrated into the overall plan. Since the offensive IO actions and desired effects are stored in the cause-and-effect network (CAEN), it would be useful to be able to represent the influence actions and desired effects in the CAEN as well. The good news is that such a mapping is, indeed, possible, and has been implemented in at least one of the modern IO planning tools, as described in Chapter 9.

All of the planning elements of the influence operation can be stored along a link on the CAEN from the desired effect to the targeted Red activity. In most cases, the targeted Red activity for Influence Operations is "protect," as the opponent is trying to "protect" their population from the messages the United States is trying to send. These planning elements include:

- The group targeted to receive the message;
- The message purpose;
- The selected change method and submethods;

- The selected delivery mechanism (and associated timing parameters);
- The MOEs, MOPs, and MOCs associated with the influence operation.

The message itself may or may not be stored on the link, as the message itself may be fairly extensive (especially if video is the delivery mechanism). However, a link to the message is useful so that the planner and plan reviewer can have ready access to that information.

In some cases, the planner is receiving some of the planning factors from a reachback center. Therefore, the request for analysis (RFA) process will provide the requested plan elements and analytic results to be stored or linked to along the link in the CAEN. (See Section 5.6 for a brief description of the RFA process.)

The end result is that the complete plan, as represented by the CAEN, includes the influence operations elements, the offensive IO elements, and the kinetic elements as part of the same plan. This mechanism allows for the nonkinetic IO options to be compared on an equal footing to the more traditional kinetic options, as well as to the influence operations components to the plan. For the first time, all of the influence, offensive IO, and kinetic effects-based operation elements can be represented and compared in a single, quantifiable format. This means that as long as there is some quantitative (even subjective quantitative) measure of how well each plan component is expected to work, the employment option COAs can be compared on an equal footing across kinetic and nonkinetic options, including influence operation options.

Overall, our approach to influence operation planning support provides a concept for a standardized PSYOP COA development, comparison, and selection at strategic, theater, and operational levels that is scalable to any echelon. It integrates PSYOP from the start with planning at the strategic/theater level. The idea battlespace concept helps scope the problem, supports situational awareness, and supports feedback and assessment. This approach is also applicable to public affairs, albeit with a reduced set of change methods and submethods appropriate for the constraints under which public affairs must operate.

The approach described defines a standard set of change methods and options, highlights the need to track the level of attention on a message, and keeps the complete set of options in front of the planner. This approach emphasizes different elements of current Joint and Service PSYOP doctrine, clearly defines the purpose of a message, adds the concept of the idea battlespace to help visualize the problem space, and introduces the concept of gaining and maintaining attention as a key element of influence operations planning. Whereas current doctrine emphasizes new message creation and delivery, the increased emphasis on a standard set of change methods and submethods gives planners a clearer understanding of what is possible and desirable in the realm of influence operations.

References

[1] Joint Publication 3-13, *Information Operations*, February 13, 2006.

[2] Air Force Doctrine Document 2-5, *Information Operations*, January 11, 2005.

[3] Joint Publication 3-53, *Doctrine for Joint Psychological Operations*, September 2003.

[4] Al-Sahhaf, Mohammed Saeed, "Iraqi Information Minister Quotes;" http://www.military-quotes.com/, April 2005.

[5] Renfro II, Robert S., and Richard F. Deckro, "A Flow Model Social Network Analysis of the Iranian Government," *Military Operations Research*, Vol. 8, No. 1, 2003, pp. 5–16.

[6] Zdon, Al, "Persian Gulf War, 10 Years Later," IWAR Web site, 2001, http://www.iwar.org.uk/psyops/resources/gulf-war/13th_psyops.htm.

[7] "Pentagon Surfing 5,000 Jihadist Web Sites," MSNBC online news, May 4, 2006, http://www.msnbc.msn.com/id/12634238/?GT1=8199.

[8] "Mixed Reaction Across Asia to Anti-WTO Violence," *CNN transcripts*, December 1, 1999, http://transcripts.cnn.com/1999/ASIANOW/east/12/01/wto.asia.reax/index.html.

[9] "Propaganda of the Deed," *Wikipedia Online Encyclopedia*, May 12, 2006, http://en.wikipedia.org/wiki/Anarchist_terrorist.

[10] *Country Reports on Terrorism, 2004*, United States Department of State, April 2005, http://library.nps.navy.mil/home/tgp/qaida.htm .

[11] Burd, Stephen, "Scientists See Big Business on the Offensive," *The Chronicle of Higher Education*, December 14, 1994, http://chronicle.com/data/articles.dir/art-41.dir/issue-16.dir/16a02601.htm.

[12] "Brown and Williamson Faces Inquiry," *The Wall Street Journal*, February 6, 1996.

[13] Leckie, Robert, *The Wars of America*, Vol. 2, New York: Harper and Row, 1968.

[14] "Tobacco Industry Undermines Public Health Efforts Around the World," *Corporate Accountability International/InFact*, http://www.infact.org/health.html, February 2006.

[15] Freedman, Alix, and Laurie P. Cohen. "Smoke and Mirrors: How Cigarette Makers Keep Health Question 'Open' Year After Year." *Wall Street Journal*, February 11, 1993.

[16] Mooney, Jake, "A Welcome Mat Withdrawn," *New York Times*, August 14, 2005; http://www.doe.org/news/pressdetail.cfm?PressID=226.

[17] Bill, James A., "Islam: Misunderstood Throughout the World," http://english.islamway.com/bindex.php?section=article&id=91, May 2006.

[18] Weisman, George, Memo delineating the tobacco industry's responses to the 1964 report "Smoking and Health: Report of the Advisory Committee to the Surgeon General of the Public Health Service," Phillip Morris, 1964.

[19] Payne, Kenneth, "The Media as an Instrument of War," *Parameters*, Spring 2005, http://carlisle-www.army.mil/usawc/Parameters/05spring/payne.htm.

[20] "Joseph Goebbels," *Wikipedia Online Encyclopedia*, May 25, 2006, http://en.wikipedia.org/wiki/Goebbels.

[21] Al-Sahhaf, Mohammed Saeed, Military, "Iraqi Information Minister Quotes," http://www.military-quotes.com/, April 2006.

[22] Parks, W. Hays, "The 1954 Hague Convention for the Protection of Cultural Property in the Event of Armed Conflict," http://www.kakrigi.net/manu/ceip4.htm.

[23] El-Magd, Nadia Abou, "Egypt Frees Al Jazeera Cairo Bureau Chief," Associated Press report appearing in the *Boston Globe*, April 27, 2006, http://www.boston.com/news/world/middleeast/articles/2006/04/27/egypt_frees_al_jazeera_airo_bureau_chief/.

[24] "Katrina Victims Blame Racism for Slow Aid," MSNBC.com, December 6, 2005, http://www.msnbc.msn.com/id/10354221/.

[25] "Falklands/Malvinas War," Global Security Web site, April 27, 2005, http://www.globalsecurity.org/military/world/war/malvinas.htm.

[26] Shulman, Eduard, "Voice of America," from *Amerika: Russian Writers View the United States*, Normal, IL: The Dalkey Archive Press, http://www.sumlitsem.org/russia/amerika/trans_shulman.htm.

[27] Hendawi, Hemza, "Iraqi Television Knocked Off the Air After US Says Allies Fire Cruise Missiles into Baghdad," Associated Press, March 25, 2003, http://www.sfgate.com/cgi-bin/article.cgi?f=/news/archive/2003/03/25/international2358EST0933.DTL.

[28] Kerchner, Capt. Phillip M., Richard F. Deckro, and Jack M. Kloeber, "Valuing Psychological Operations," *Military Operations Research Journal*, Vol. 6, No. 2, 2001.

[29] Pew Institute, "War with Iraq Further Divides Global Politics," *Views of a Changing World 2003*, report released June 3, 2003, http://people-press.org/reports/display.php3?ReportID=185.

[30] Rouse, Ed, "The Gulf War," Psywarrior.com, http://www.psywarrior.com/gulfwar.html, February 2006.

7

Planning Methodologies and Technologies for Defense Missions in IO

> *In a general sense, camouflage is the art of concealing the fact that you are concealing.*
> —Guy Hartcup, *Camouflage: A History of Concealment and Deception in War*

This chapter focuses on defensive information operations, which include the core capabilities of computer network defense (CND) of CNO, Electronic Protection (EP) of EW, Operations Security (OPSEC), and Military Deception (MILDEC), and the supporting capabilities of information assurance (IA), physical security, and counter-intelligence (see Figure 4.1). As mentioned in the previous chapter, this book categorizes military deception, operations security, and computer network defense as three defensive types of information operations. According to Joint doctrine (JP 3-13), military deception, operations security, and computer network operations are three of the five core capabilities of IO. In contrast, U.S. Air Force Doctrine places PSYOP, military deception, and operations security in the influence operations bin.[1]

---

1. Note that while electronic protection (EP) is assumed under this book's defense category, we do not highlight which parts are EP in the examples. Note that the OPSEC examples below have many EP elements. In general, EW tends to be focused on the tactical level for both implementation and effect, but detected friendly transmissions could have an operational-level effect. Chapter 9 also mentions an EW planning tool called "Builder" that is currently used in planning.

Section 7.1 describes the general purpose of defensive IO planning, and how it differs from offensive IO and influence operations planning. Section 7.2 presents our methodology for planning military deception (MILDEC) operations and how MILDEC planning leverages the offensive and influence planning processes and products. Section 7.3 presents operations security (OPSEC) planning, how it also leverages the offensive and influence processes and products, and how it relates to military deception planning. Measurements of the degree of success of the MilDec plan and OPSEC plan are closely intertwined. Section 7.4 presents planning principles and techniques for computer network defense (CND).

7.1 How Defense Differs from Attack and Influence

Defensive IO is, by nature, reactive rather than proactive. In all defensive operations, the initiative is ceded to the opponent in any area in which you choose to defend. As a result, defensive IO consists mostly of setting up objectives and contingencies to *detect and react* to anticipated events, and then monitoring the situation to determine *when and how* you should respond based on your plan.

Defensive IO is not entirely passive, however, just as traditional military defense is not entirely passive. For traditional land warfare defensive operations, patrolling is used to keep tabs on enemy activities, collect intelligence, provide early warning, protect forces from enemy observation, and disrupt enemy preparations. Comparable activities are available to the defensive IO planner. For example, taking out, disabling, or denying access to enemy airborne sensors is a proactive action in support of OPSEC to preclude enemy aerial observation. Transmitting false radio signals to distract enemy SIGINT sensors and commanders is a proactive action in support of military deception. Setting up a honey pot, or false part of the network to attract and track potential attackers, is a proactive step in computer network defense.

The primary reason that this book categorizes MILDEC, OPSEC, and CND as defensive IO rather than as influence operations is that the ultimate purpose of military deception, OPSEC, and CND is to protect your assets from enemy observation and interference, and to distract or deceive potential enemy observers and decision-makers from your actual intent and objectives. In each case, there are things you either want the enemy to see or not see, depending on your overall plan, and things that you want to see to measure the status of your plan. The key to good defensive IO planning is to clearly define what you want the enemy to see or not see, and define how and when you will measure the degree of success in achieving those desired results. Moreover, preplanned responses or contingency plans help ensure that the response is sufficiently rapid and well planned to be effective and that the response does more good than harm to the defensive side. This is the basis of defensive IO planning and the key to the measures necessary to monitor progress in defensive IO. Just as there are MOPs, MOEs, and MOCs in influence operations, similar types of measures are used by defensive IO planners.

A secondary reason that we categorize MILDEC, OPSEC, and CND as part of defensive IO is that these planning processes and products are more similar to other areas of IO defense than the planning processes and products for influence operations. Since the methodology for these three types of IO defense are more similar to each other than to influence operations, we place them in the same bin. This chapter will show how MILDEC and OPSEC planning can be measurable, reproducible, and auditable.

7.2 Military Deception Planning

Military deception consists of both art and science. The *art* involves perceiving the world through the eyes of the opponent to help determine what will fool him. Successful military deception depends on the ability to identify the predisposition of the opponent to believe something and exploiting that predisposition to distract the enemy from the real threat. The art of military deception involves a great deal of imagination and understanding of the opponent's way of thinking.

The *science* of military deception tries to codify methods and measurables that have tended to be successful in previous military deception operations. For example, a common strategy in successful military deception is recognizing the fact that one's opponent often relies upon his predisposed view of the world. The ability to accurately identify the enemy's predisposition is still an art, but the methodology to first try to identify the enemy's predisposition is part of the science. This section presents seven steps to describe the science portion of military deception planning. Before presenting these seven steps, however, some background information is presented.

As mentioned in Chapter 5, in attack planning, the friendly courses of action are compared, and the best course of action is selected. Unfortunately, the enemy's predisposition may be to respond to the best friendly course of action, which creates a quandary for the friendly commander. If the best course of action is the one the enemy expects, is it really the best course of action? Considering the enemy's intentions and perceptions is also a good way to choose the best course of action. As Sir Winston Churchill said [1], "No matter how enmeshed a commander becomes in the elaboration of his own thoughts, it is sometimes necessary to take the enemy into account," that is, one must consider the enemy when making one's plans.

Confederate General William Bedford Forrest said, "Hit 'em where they ain't," while Sun Tzu said [2], "So in war, the way is to avoid what is strong and to strike at what is weak." If the enemy knows where you are going to attack, he will have his forces ready at that location. But since the enemy will "orient" against the friendly action he expects, he will not be properly oriented to respond to the action he does not expect and will be "disoriented" in his response. Therefore, it is usually more cost-effective to take the less expected course of action, and therefore the best COA is the one the enemy does not expect.

For example, the Allied invasion of Axis-occupied France in 1944 recognized that the shortest crossing point was at Calais, and a successful invasion at Calais would threaten to cut off most of the German army in France. Since the Germans recognized this threat, they oriented their forces, and especially their armored reserves, against a Calais invasion. The Allies recognized the risk of invading where the Germans were the strongest, and therefore selected Normandy for the landing site, and used the Calais landing as the basis for their deception operation. The details of this Calais deception operation were extensive, with a phony Army organization headed by General Patton, dummy bivouac areas, recorded tank movements played on loudspeakers, and false radio signals [3, 4].

In a similar manner, the Coalition during Operation Desert Storm considered an amphibious landing from the Persian Gulf as one option to outflank the defending Iraqis facing south. However, the Iraqis had also read history, and the Inchon landing of the Korean War played heavily on their minds. Subsequently, Coalition intelligence reports showed that all of the likely landing locations were heavily mined and fortified by the Iraqis, who had also placed a large number of troops facing toward the Gulf. Since the Iraqis seemed to be anticipating the threat from the sea, the Coalition decided to focus on an overland attack as the main thrust. The original intent was to advance directly from the south, but an alternative plan developed by the "Jedi Knights" raised the possibility of the famous "left hook" through trackless desert [5]. Unbeknownst to the Coalition planners at the time, the Iraqis had assumed that no military action could be taken through that expanse because everyone who trained in that area had gotten lost. So there were actually two elements to the deception plan in the First Gulf War: the preground war show of force of Coalition amphibious capability to keep Iraqi forces facing the sea, and the initial ground assault directly from the south, which was a feint to keep the Iraqis looking south and not to the west. When the Coalition liberated Kuwait City, they found that the Iraqi III Corps commander's sand table showed virtually all Coalition avenues of approach coming from the sea, which indicated a successful deception operation [6].

In Chapter 5, we described the need to keep an audit trail of all planning decisions, including the courses of action that were rejected and why. One advantage of this approach is that the best military deception plan is likely to be found in one of the rejected COAs. The military deception plan most likely to succeed against the enemy is one that is credible. A COA that has been elaborated to some detail, evaluated, and rejected, is thus a good candidate to become the basis of the military deception plan. Such a COA would be credible to the planning side simply because it was seriously considered. The enemy also may be predisposed to respond to that COA; indeed, that COA may have been rejected simply because the enemy *was* predisposed to it. In other words, the plan that we are not using is the one that we are likely to be trying to convince the enemy that we *are* using.

Therefore, the first step in military deception planning is to identify the friendly COA that the enemy is predisposed to assume will be our actual plan. The military deception plan is developed based on the friendly COA against which the enemy has decided, or may decide, to orient its forces. Making this identification is easier said than done, partly because one must also consider enemy deception in the disposition of its forces. Still, the evidence for potential enemy deception is also considered when trying to select the best friendly COA to be executed.

Since the rejected COA has already been partially prepared, taking the rejected COA and turning it into the military deception plan for this operation has a solid starting point in terms of credible elements. This can be seen by viewing the cause-and-effect network described in Chapter 5 and by reviewing the objectives, tasks, and actions considered in that COA. These objectives, tasks, and actions are compared to those for the COA that was selected to be the plan, and differences and similarities between them are identified. What the military deception planner is looking for at this stage are those features of the military deception plan that are different from the actual plan, and to identify those features that, if accepted by the enemy, could cause the enemy to believe the military deception plan is the one being prepared for execution.

Therefore, the second step of MILDEC planning is to determine the features of the military deception plan that are different from the actual plan in order to define which features of the deception plan we want the enemy to see (as opposed to details of the actual plan that we do not want the enemy to see). Note that if the enemy is never allowed to see selected features of the military deception plan, then the enemy might not be convinced and might opt for a more flexible posture instead of committing to defend against the deception plan.

The third step is to define the signatures and indicators of the features of the military deception plan we want the enemy to see. For example, Figure 7.1 shows sample elements of a military deception plan. Note that these plan elements are features of deception that we want the enemy to perceive so that the enemy believes that the deception plan is the actual plan. Planning for the features we want to hide is addressed in Section 7.3; however, the MILDEC feature planning and the OPSEC feature planning should be performed in a coordinated manner and not in isolation.

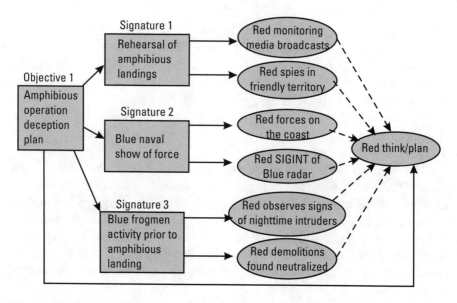

Figure 7.1 Sample elements of a military deception plan.

The fourth step is to define how we think the enemy might be able to detect the signatures and indicators we want them to detect. The first column of Table 7.1 shows what we want the enemy to see, while the second column shows *how* we expect the enemy to detect them.

For example, during Operation Desert Storm, we wanted the enemy to observe our amphibious landing rehearsals, but did not want them to see the buildup of forces in the Western Desert. So we selected mechanisms by which they could observe the preamphibious activities without allowing the Iraqis to have airborne reconnaissance. To support this deception, Coalition naval forces performed show-of-force maneuvers at sea, while frogmen performed actions at night whose results were intended to be detected during the day. In addition, the press was allowed to observe Marine landing rehearsals, because the Iraqis were watching international news as a source of preground-war intelligence.

The fifth step is to define the methods by which we will *measure* the achievement of enemy detection of the signatures and indicators of the features of the deception plan. These measures can be based on reasoning, such as "no enemy sensors are flying, so there can be no low-level aerial detection," or they could rely on other intelligence sources to corroborate the detection.

The sixth step occurs during execution monitoring of the plan. (See Chapter 8.) The objective of step six is to confirm whether the enemy has actually detected what we wanted them to detect of the deception plan and determine whether the enemy has accepted the deception plan as our actual plan. Execution monitoring of the deception plan involves tracking which features of the

Table 7.1
Sample Signatures and Indicators of Enemy Detection for MILDEC

Deception plan is an amphibious operation	Red sensors that could detect signature
Signature 1: Rehearsal of amphibious landings	Red monitoring of media broadcasts
	Red sympathizers or spies in friendly territory
Signature 2: Naval show of force	Red forces on the coast
	Red SIGINT of Blue radar operating offshore
Signature 3: Blue frogmen activity preparatory to amphibious landing	Red forces observe signs of nighttime intruders in daytime
	Red engineers note that demolitions neutralized

deception plan we believe the enemy to have seen and which detections the enemy appears to have accepted as true. Such confirmation is not always easy, but using modern intelligence capabilities, there are multiple ways by which the enemy detection and acceptance of deception plan features may be corroborated. In some cases, the detection may be confirmed, while in other cases, the acceptance may not be confirmed. In either case, the MILDEC planner and warfighter need to provide an estimate of the probability that the enemy has accepted the deception plan as the actual plan, and provide a basis of evidence for that estimate.

The seventh and last step of military deception planning is to compare the confirmed and likely enemy detection and acceptance of features of the MILDEC plan with any features of the actual plan that leaked through the OPSEC plan. The details of this comparison will be described in Section 7.3, but the following point is the key to the successful military deception plan: *The execution of the military deception plan does not have to be perfect—it just has to be better than any failures of the OPSEC plan.*

To summarize the seven steps of military deception planning:

1. Identify the friendly COA that the enemy may be predisposed to assume as the basis for the deception plan.
2. Determine the features of the military deception plan that are different from the actual plan to determine which features of the deception plan that we *want* the enemy to see.
3. Define the *signatures and indicators* of the features of the military deception plan that we want them to see.
4. Define the methods by which the enemy will detect these signatures and indicators and the actions that can be taken to ensure the enemy sees the signatures and indicators that we want them to see.

5. Define how we will *measure the achievement* of enemy detection and acceptance of the signatures and indicators of the features of the deception plan.

6. During deception plan execution, attempt to confirm via various intelligence channels whether the enemy detected the signatures we wanted them to detect, and whether the enemy appears to believe those indicators. The measurement of the success of the deception plan can therefore be quantifiable, even if subjectively quantified by the deception planners and warfighters.

7. Compare what we think the enemy saw of the deception plan to what we think the enemy saw of the real plan that leaked through the OPSEC plan. The execution of the military deception plan does not have to be perfect; it just as to be better than any features of the real plan that leaked through the OPSEC plan.

7.3 Operations Security Planning

In this book, the methodology for OPSEC planning is parallel to the methodology for military deception planning and also has seven planning steps. What is different is that for OPSEC planning, the emphasis is on the features of the actual plan that we do not want the enemy to see and how we plan to measure the achievement of the OPSEC plan by itself and in comparison with the MILDEC plan. In other words, OPSEC planning focuses on what we do not want the other side to see and whether we think or know the other side has seen it and accepted it. (This is where electronic protection provides its greatest contribution.)

Since OPSEC planning starts with the COA selected to be the plan to be executed, there is no decision to be made as to which COA to select. Therefore, as the first step of OPSEC planning, the OPSEC planner starts with the actual plan, which may already be in the cause-and-effect network format.

The second step is to identify the features of the actual plan that make it distinct and uniquely identifiable from the deception plan. Features that are the same for both the actual plan and the deception plan do not contribute to the enemy distinguishing between the two plans, and therefore are of less interest to the OPSEC or MILDEC planner. For example, if supplies and reinforcements are planned to come in at a certain airport in both the actual plan and the deception plan, then there is less need to hide the fact that supplies and reinforcements are arriving at that airport. Basic tactical force protection, physical security, and electronic protection procedures are good for every force, regardless of whether their actions are in support of the actual plan or the deception plan. Operational-level security planning, however, needs to focus on those aspects of the actual plan that, if detected by the enemy, would jeopardize the whole operation.

No one has the time or resources necessary to protect every operational aspect of the actual plan. Instead, the OPSEC planner needs to focus on the features that, if detected, would cause the enemy to distinguish the actual plan from the deception plan and determine that the actual plan is the one we intend to use. This defines the set of features of the actual plan that we do not want the enemy to see.

The third step is to define the signatures and indicators of the features of the actual plan that we need to hide from the enemy. This list of signatures and indicators should be prioritized, so that the most important signatures and indicators are high on the list of those that need to be hidden from the enemy. Figure 7.2 provides an example of some signatures and indicators that would need to remain hidden from the enemy.

It may also be useful to have some elements of the OPSEC plan implemented before some elements of the military deception plan, or vice versa. For example, ensuring radio silence under OPSEC before launching phony signals from the military deception plan is a likely combination of steps. This means that the *synchronization* of the military deception plan and OPSEC plans should be done by a single planning cell to ensure that synergistic plan elements can be coordinated, and conflicting ones avoided. This is further described in Section 7.3.1, along with some classification considerations.

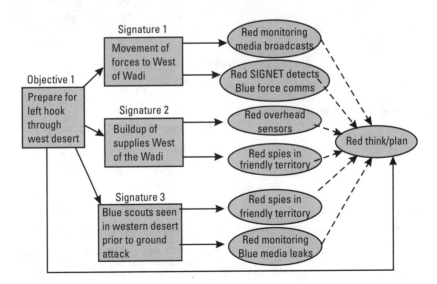

Figure 7.2 Sample elements of an operations security plan.

The fourth step is to define the *methods* by which the enemy is likely to detect the signatures and indicators that we do not want them to detect and how these detections can be blocked. The first column of Table 7.2 shows the signatures we want to block, while the second column shows the types of enemy sensors that could detect each signature.

Using reasoning and information on enemy sensor capabilities, the OPSEC planner should be able to determine which features of the actual plan could be detected by the enemy, and how it could be blocked by our forces. For example, if enemy aircraft could observe key features of the actual plan, then the OPSEC plan needs to determine how the enemy sensors could be precluded from flying, or cause them to otherwise not see that feature even if the sensor is flying. In a similar manner, if Red can monitor Blue radio communications, then radio discipline is essential to the undetected movement of forces.

The fifth step is to define how we will *measure* whether or not the enemy actually accomplished those detections. The planner defines how we might measure whether or not each enemy sensor could or could not detect the signatures we want to block. This definition of measures helps define requests for information that will be performed during plan execution.

The sixth step is performed during plan execution, where the OPSEC planner and warfighter attempt to determine whether the enemy might have detected any signature they did not want the enemy to detect. (See Chapter 8.) Using various intelligence capabilities, the OPSEC planner and warfighter receive reports that indicate whether a specific feature of the plan was or was not detected by the enemy, and whether the enemy believed that detection. Table 7.2 provides an example of the types of intelligence information that can help determine whether the OPSEC plan is protecting the features of the

Table 7.2
Sample Signatures and Indicators of Enemy Detection for OPSEC

Actual plan is a left hook through the desert	Red sensors that could detect signature
Signature 1: Movement of forces to the west of the Wadi	Red monitoring of media broadcasts
	Red SIGINT detection of Blue force's communications
	Red sympathizers or spies in friendly territory
Signature 2: Buildup of supplies to the west of the Wadi	Red overhead sensors (aircraft or satellite photos)
	Red sympathizers or spies in friendly territory
Signature 3: Blue scouts in Western Desert prior to ground assault	Red forces patrolling Western Desert
	Red monitoring for Blue media leaks

actual plan from enemy observation. For example, if no enemy aircraft are flying, then blocking enemy observation by aircraft can be confirmed. If the enemy has a radio signal direction finder, but no friendly radios are transmitting within range, it is probable that the enemy did not detect any friendly signals. Conversely, if we cannot confirm the location of enemy SIGINT and that sensors and friendly forces have been recently transmitting, then it is possible that there was a leak that the enemy might have detected. Through HUMINT and other methods, it might be possible to determine whether the enemy believed what he saw, even if we confirm he has actually seen it. This leads to the last step.

The seventh step is to compare what we think the enemy observed of the actual plan and compare it to what we think the enemy saw of the deception plan, and determine which perception has been accepted by the enemy. It may be possible to provide a quantitative, even if subjective, estimate of the probability of the enemy believing the deception plan versus having discerned the actual plan. Once again, intelligence confirmation of the belief would be preferable, but is not always possible. Therefore, the OPSEC planner and warfighter will generally need to rely on a reasoned probability of success if such independent confirmation of enemy belief is not possible. As before, the execution of the OPSEC plan does not have to be perfect, only better than what the deception plan has successfully presented to the enemy.

Like military deception, OPSEC planning is both an art and a science, but it is more science than art when compared to military deception planning. There are standard operating procedures for tactical-level OPSEC, and the same principles tend to apply to operational-level OPSEC. Even so, imagination plays an important role in OPSEC, because the OPSEC planner needs to be able to think through the threats to the detection of the real plan through the mind and eyes of the opponent.

To summarize the seven steps of operations security planning:

1. Begin with the COA selected to be the plan as the basis for the OPSEC plan.
2. Determine the features of the actual plan that are different from the deception plan to determine what features of the actual plan that we do not want the enemy to see.
3. Define the signatures and indicators of the features of the actual plan that we do not want the enemy to see.
4. Define the methods by which the enemy will detect these signatures and indicators. This involves identifying enemy sensor types, determining how they could detect what we do not want them to detect, and determining how to best block that potential detection opportunity.

5. Define how we will measure the achievement of possible, probable, or known enemy detection of the signatures and indicators of the features of the actual plan.

6. During actual plan execution, attempt to confirm via various intelligence channels whether the enemy detected the signatures we did not want them to detect and whether the enemy appears to believe or have accepted these indicators. The measurement of the success of the OPSEC plan can therefore be quantifiable, even if subjectively quantified by the OPSEC planners and warfighters.

7. Compare what we think the enemy saw of the actual plan that leaked through the OPSEC plan and compare it to what we think the enemy saw and believed of the military deception plan. The execution of the OPSEC plan does not have to be perfect; it just has to have fewer leaks than the features of the deception plan accepted by the enemy.

7.3.1 Additional Issues in MILDEC and OPSEC Planning

Military deception and operations security planning are currently performed by different cells. Part of the reason is that the military deception plan is very fragile in that it is quickly negated if the enemy figures out that it is a deception. As a result, the MILDEC plan is highly classified and planned by a small, compartmentalized planning cell. By design, few personnel on the planning staff have access to the military deception plan.

In contrast, every headquarters plans for OPSEC in some way. OPSEC is everyone's responsibility [7], and as a result everyone knows his or her local OPSEC plan. This separation of planning between military deception and OPSEC planning cells means that the *creation of the synergy* between the military deception and OPSEC plans must be performed by the military deception planning cell. Organizationally, that is the only cell that has all of the information necessary to define the features that distinguish the actual plan from the deception plan. Therefore, at the operational level, the military deception planning cell needs to pass to the OPSEC planning cell the list of key features to be blocked by the OPSEC plan. This provides a set of OPSEC requirements that prioritizes the OPSEC planning efforts, in addition to the standard force protection tactical OPSEC plans.

In a similar manner, when the actual plan is being executed, the *comparison* of measures between the levels of achievement of the deception plan and the actual plan must be performed by the military deception planning cell. While the OPSEC planners and warfighters will have access to the measures of achievement of the OPSEC plan blocking enemy observation of the actual plan, only the MILDEC planning cell will have access to the other half of the data necessary to make the comparison to estimate whether the enemy believes the deception plan or has detected the actual plan.

There is also an important time consideration. The military deception plan does not have to be successful forever—it only has to be successful long enough for the enemy to be unable to react effectively against the actual plan. Once again, the intent is not for perfection, but only for the deception to be good enough long enough to achieve the desired effect.

Another responsibility for operations security and military deception planning is how to handle mistakes. For example, if there is a leak of a feature of the actual plan, there needs to be a way to explain it away or otherwise discount it so that the enemy doubts that the leaked information is valid. It is better to have a plan for how to handle these leaks before they occur, but sometimes they are so amazingly huge leaks as to defy belief. In such cases, it might be better to try and convince observers that the leak was part of a deception plan rather than part of the real plan.

For example, a massive security leak occurred just before the invasion of the Low Countries and France in 1940, when a German officer's attaché was captured with the detailed invasion plans. The German two-man aircraft had landed in Belgium, and the Belgians believed the captured plans for Germany's invasion of Holland and Belgium to be authentic. However, the British and French erroneously believed that the documents were planted and therefore dismissed the information [8]. Thus, even in the case of a serious leak through the OPSEC plan, the actual plan might be protected not only by the enemy's reaction, but also by hints dropped by the side that suffered the leak.

In a similar manner, a central responsibility of military deception is to ensure that the other side does not stumble across obvious deception activities—unless intended. Therefore, even if mistakes in military deception planning and execution occur, the actual plan and other parts of the deception plan can be used to advantage even if part of the deception plan has been discovered as being a deception. For example, during World War II, the Japanese built a runway on an island with dummy aircraft parked at the end of the runway. Since the dummy aircraft construction was not very convincing, the American pilots made it a point of pride to not bomb the dummy aircraft at the end of the runway. One day, however, a couple of bombs landed long, blowing up not only the dummy aircraft, but also generating a large secondary explosion! Apparently, the Japanese had placed their fuel dump under the dummy aircraft, knowing that they would not be intentionally bombed.[2] In a similar manner,

---

2. The author could not find a public reference for this specific anecdote, but did find a reference to the Japanese being famous for the "double bluff." In a double bluff [9], "decoy aircraft were given 'a childish kind of camouflage' causing them to be "mistaken for true works which have been insufficiently camouflaged."

even elements of the deception plan that are known to have been discovered can be used to further the goals of the actual plan.

Overall, military deception involves defining what the enemy should see of the deception plan, while operational-level OPSEC involves what the enemy should not see of the actual plan. This approach allows the MILDEC and OPSEC planners to define key features of each plan that must be shown or hidden and estimate how the enemy might be able to detect features of the deception plan and the actual plan. The MILDEC and OPSEC planners can define measures for the level of achievement of each plan, and can try to confirm the level of achievement through corroborating intelligence sources. Lastly, the military deception plan does not have to be perfect; it just has to be better than any detections of the actual plan that leaked through the OPSEC plan.

7.4 Computer Network Defense Planning

In most fields of competition, the advantage shifts back and forth between attack and defense due to new technology, or improved methods of using existing technology. The rifle gave the tactical advantage to the defender in the American Civil War. The machine gun gave an even greater advantage to the defender in World War I. Between the World Wars, while France was building the Maginot Line, the Germans were refining the blitzkrieg, adding armor to the Sturmmann (storm trooper) tactics to give the advantage back to the attacker [10]. Nuclear weapons gave a huge degree of advantage to the attacker, especially in terms of first strike.

In computer network operations, the attacker currently has the advantage. Today's heavy reliance on modern information technologies has led to significant vulnerabilities in both civilian infrastructures and military endeavors (as described in Chapter 2). Besides computer networks, most other networks in the infrastructure are computerized or have computerized components. As a result, computer network attacks can be applied to almost any type of network with electronic components. Since the public and commercial emphasis was on providing ubiquitous connectivity, much less emphasis was given to security. As a result, a number of inherent vulnerabilities give the attacker the advantage in the cyber-attack cyber-defense competition. This advantage will decrease over time, and may eventually favor the computer network defender for a period of time. However, at the time of this writing, the computer network attacker still has the advantage over the computer network defender.

The network defender is currently in reaction mode. Like most defensive operations, the initiative has been forfeited to the attacker. As a result, the network defender is usually setting policy, defining a defensive posture, and monitoring whether the policy has been properly implemented and when threats start appearing. The role of the defender is to identify these threats and respond to them quickly and effectively. The main problem with identifying threats is determining the distinction between a threat versus normal operations, as well as identifying the source and intent of the threat. The main problems with responding is that the response needs to do more good than harm to your network, needs to effectively identify and block enemy intent, and, if any sort of active response against the perceived attacker is used, needs to make sure that response affects only the correct target. We will address each of these issues in more detail next.

7.4.1 Network Defense Planning Principles

Events often happen quickly in networks. Due to the interconnectivity and extensive capabilities in most networks, the effects of a given threat can rapidly become significant. The speed at which threats can complete the cycle from insertion to completion of intended effect is very fast. For example, in experiments with unprotected computers connected to the Internet, compromise of the computer occurs in only a few minutes [11]. As a result, a network defender needs to have predefined responses to these rapidly evolving threats. The network defender does not have a lot of time to think about what to do if a threat occurs. Once the threat has been detected, the response needs to be rapid or else it will be too late.

Since the network defender is almost always reacting to threats, the first principle of network defense planning is to *prepare different forms of contingency plans*. There are many different types of contingency plans, from automated to manual, from near term to long term, and from individual machines to complete networks.

The only completely secure network is the one that no one can access. Although this may seem obvious, this fact sometimes comes back to haunt those trying to increase their level of security. There are tradeoffs associated with increased security, such as reduced access by those who should have access, reduced bandwidth, and a reduction in just about every other measure of network performance. Anything that the network defender can do to better defend the network usually results in reduced network performance.

This leads to our second principle in planning network defense: When responding to a threat, *do less harm than good* to your own network. This is similar to the part of the Hippocratic oath doctors take to do no harm. As a network defender, you might reduce some performance of the network, but if the cure is worse than the threat, then you are actually causing more damage to the network than does the threat itself.

In 1999, my colleague Randy Jackson and I developed a linear program to find the optimal solution to the problem of which defensive responses provided the best protection given the damage the response would do to network performance. Today, measuring the impact of defensive responses on a network is easier than ever. Besides having a test network with which to run the defense options, there are now models optimized to scan networks and predict the effect of policy changes. For example, Skybox Security has software that collects the network data directly from the network on a secure channel, and allows the network defender to see the effect that policy changes will have on the connectivity required to keep the organization running. While the Skybox software does not provide network performance prediction data, other simulations can, such as OPNET from OPNET Technologies.

Based on the availability of such tools, a network defender can simulate the effects of different responses on the performance of their defended network. Even without these software tools, the network defender can experiment with the effects of responses during off-peak hours. In either case, the network defender should not be applying defense responses when the effects of those responses on the overall performance of the network are unknown. In such a case, the damage done to the network and the operations of those who rely on the network could be catastrophic.

One early policy approach to network defense policy was setting the information condition (INFOCON) levels. INFOCON was the initial effort to both categorize the threat and implement a predefined set of defensive actions or postures to quickly respond to that threat. Its role model was DEFCON (Defense Condition), a Nuclear Age method to define what standard defensive steps would be taken when the specified threat level was deemed to exist [12]. INFOCON defined five levels of defensive posture (normal, and alpha through delta) that included an associated set of defensive postures to respond to each perceived threat level [13]. The problem was that the required INFOCON response was network-wide. As a result, the effects of going to INFOCON "charlie" could disrupt normal operations so badly that only the most significant threat would cause decision-makers to go to that level of reaction. According to a 2001 GAO Report [14], "Inexperienced personnel may overreact and implement drastic countermeasures, resulting in self-inflicted problems, such as degraded system performance or communication disruptions."

This leads to our third principle in planning network defense: *flexibility in response*. In order to be effective, the response must be tailorable to the threat. In particular, only the threatened portions of the network need to shift to a higher defensive posture, letting the rest of the network operate at higher levels of performance. One such tool, the Integrated Information Infrastructure Defense System (I3DS) allows the INFOCON to be set at the host-machine level. Like Cisco's Security Agent and Symantec's Internet Security, the system administrator presets the software, processes, and ports allowed to be used on a given host. However, I3DS provides flexibility by allowing the network defender to quickly modify what is allowed at each host by raising the INFOCON only for that machine. This gives a substantially greater degree of flexibility to the network defender.

Defining rules regarding what is allowable per machine has advantages and disadvantages. The advantage is that it provides a great deal of flexibility of response to the network defender. This works particularly well when there is great similarity as to what is allowed among the host machines on a network. However, when there is a wide variety in the software, processes, and ports allowed per machine, then the process of defining, monitoring, and managing the network could potentially become a network management nightmare.

For example, it is more efficient to remotely install new security configurations on all the machines on a network. However, what if a machine on the network is already compromised? By placing the security configuration in an already compromised machine, the fox is already in the henhouse and is now a trusted member of the network. The best way to confirm that every host machine on the network is authorized to be there is to manually install the security configuration on every machine on the network. Due to the large number of electronic devices already connected to mature networks, this can be a monumental task. However, it only needs to be done once, since the security upgrades can be performed remotely.[3]

Of course, no network defense software will be perfect. Someone will always come up with something that will cause a threat to remain unrecognized. How many science fiction stories have revolved around our hero getting the magic software or command into the evil empire's network, thereby causing the enemy's network to cease functioning and saving Earth once again? Unfortunately, it is not all fiction. The U.S Army is moving toward the Future Combat System, which is a mix of both manned and unmanned vehicles, networked together and in constant contact with each other. This provides a significant advantage to the FCS-equipped force, since it can operate in a network-centric fashion and beat forces much larger than itself. The problem is that all of the advantages of the network-centric force rely on the continuous or near-continuous functioning of the backbone network.

Recognizing this as a potentially significant vulnerability, the U.S. Army asked the Defense Advanced Research Projects Agency (DARPA) to examine ways to better protect wireless, ad hoc, reconfigurable networks. The Defense Against Cyber Attacks in Mobile Ad Hoc Network Systems project (called DCA MANET) is comparing design alternatives to protect critical wireless networks like FCS [15]. One of their guiding design assumptions is to configure the

---

3. Even so, new techniques are now being developed for how to install security software on already compromised machines, depending on the degree of compromise.

network to assume that it will be compromised at some point and to design the network to identify and respond to an unknown threat.

This introduces the fourth principle of network defense planning: *Have a plan for how to handle surprises*. This is not dissimilar to the information assurance principle of having a remediation or reconstitution plan after a catastrophic failure, except that the surprise response plan is executed while operations are underway. Given the severity of the surprise response plan, significant damage to the performance of the network will occur. However, this is the fallback plan. For all intents and purposes, one has called retreat and is giving up a significant portion of the virtual terrain in the network. However, having a surprise response plan is necessary to ensure that the core of the network can survive and be reconstituted for continued operations, albeit at significantly reduced capability.

Note that all of the preceding response examples have addressed how to respond within the defended network. One more response option that needs to be addressed is whether and how to respond outside the network. In one of the most public cyber-conflicts, Palestinian hackers and their supporters hacked Israeli Web sites and their supporters and vice versa. The conflict spread into several other nations due to the desire to seek vulnerable cyber-targets. A similar conflict occurred during the EP3 incident between U.S. and Chinese hackers and their supporters [16].

Since the participant hackers on all sides of these cyber-conflicts were apparently not sponsored by any nation, their concern for collateral damage was not great. However, a national or military response to a hack carries significant responsibility to avoid collateral damage. If a national or military organization did hack back, and the affected network was not where the hacker was located, it could create an international incident or political scandal. For example, if a hacker attacked via a compromised network at Amnesty International, and someone damaged the Amnesty International network by hacking back, the political fallout would be tremendous.

As a result, the preference for nations is a physical response to hacking, such as arresting hackers at home or overseas. In these cases, the identification of the perpetrator is often gleaned by experts who dissect the code, figure out styles and sources, compare event details, and profile the perpetrator.

Other physical aspects of active defense include identifying and locating malicious code on unsuspecting machines. For example, the FBI caught the perpetrators and cleaned "zombie" software off hundreds of machines that were being prepared to launch a potentially disabling denial-of-service attack against the Internet on New Year's Eve, 2000 [17, 18]. There was also a recent proposal by the Congressional Research Service to make owners of machines used to attack other machines at least partially liable for any damages as a way to provide incentives to computer owners to use better security practices [19]. So there are some active defense measures and policies that can be applied to network defense. Even so, the hack-back option is not one the United States or its military is likely to use until the technology and practice exist to provide certainty as to who the attackers are and the location from which they are operating.

Identification of the source of a threat is not just a problem in terms of who but also in terms of what. Due to the large quantity of legitimate actions occurring on a network at any given time, it is often difficult to distinguish between normal network activities and malicious activities. More professional hackers tend to use low probability of detection (LPD) attacks, such as passive sniffing, low-frequency active probing, and other nonobvious actions that fall into the noise of normal network traffic. State-sponsored network attacks tend to be better equipped and more professional. While hackers like the notoriety of a successful hack, state-sponsored professionals prefer that their activities are never discovered.

The usual solution for identifying threats is some sort of intrusion detection system (IDS) software that dutifully reports all activities on the perimeter of the network that could be considered a possible threat. In demonstrations of this software, the salesperson gleefully shows how every possible type of threat is reported in a rapid and unending stream of alerts to the network defender. The problem is that there is a high false-alarm rate. In most cases, even normal network traffic can be reported as a possible threat because an attacker could use that method as a precursor to something more significant. So the network defender is swamped in a deluge of alerts that he or she never has time to investigate.

This leads to the fifth principle of network defense planning: *Flexibility in visibility*. The network defender needs tools and techniques that allow the network defender to interactively define the amount and type of information provided. Most IDS software provides some score of relative importance of alerts, but there are nearly always more high importance alerts than the network defender can understand, let alone analyze. One can turn off the alert system, but there is no way to effectively tailor it. In other words, the network defender's visibility into the network is an all-or-nothing decision. Either you receive no information about the network, or you receive too much information about the network.

Therefore, the visibility into the network needs to be selectively tailorable by the network defender, just as the response has to be tailorable in order to be effective. Although more of an execution monitoring issue than a planning issue, the planning part involves the starting level of visibility selected for different parts of the network, and the policies for when and how the level of visibility needs to be increased.

The flexibility in visibility is especially important in addressing the problem of the insider threat. The greatest damage can and has been done by the insider rather than an external intruder due to both the level of access and the amount of time available to circumvent the security and do the damage. This is one of the main reasons why IDS software by itself is woefully inadequate to network defense. The insider is already within the perimeter, so an IDS is useless against an insider threat.

The solution is for the network defender to provide *defense in depth*, which is the sixth network defense principle. Defense in depth is the only way currently known to counter the insider threat, as well as external threats that already succeeded in penetrating the perimeter. Research into good defense in depth approaches is still underway. For example, DISA released a request for information in May 2005 for research on host-based cyber-defense software [20]. The intent is to provide a level of protection at the individual machine level that is far beyond personal firewalls and is designed to ensure that the insider cannot circumvent the security system. As another example, Symantec Internet Security is very good at blocking external threats against a machine, but all the insider has to do is turn it off, perform the malicious action, and then turn the software back on again. Most personal firewalls were not developed with the insider threat in mind.

The seventh and last network defense principle is the need to *identify and block attacker intent*. This is not an easy task, especially since even distinguishing the presence of a threat is so difficult using current network defense technologies. Assuming one was able to identify a specific threat, how does one identify the intent of the attacker, and how should one block it? Honey pots have provided some insight into this problem. A honey pot is [21]: "a computer system on the Internet that is expressly set up to attract and 'trap' people who attempt to penetrate other people's computer systems." While honey pots are good at studying the general intent of intruders, they do not help much when an intruder is actually trying to penetrate the network you are trying to defend. Even so, intruders tend to follow patterns, and patterns can be used to predict subsequent steps in an intrusion attempt. By having tools that help predict the next likely step in an intruder's attack, one can better focus the increased visibility and tailor the response to the next likely step in the attack sequence.

Effective and efficient intrusion prediction software is still in its infancy. Although pattern-matching techniques have been used to identify attempted intrusions, they are time-consuming to create and adapt to new threats. More research needs to be done in the area of predictive response planning for computer network defense.

Overall, the attacker currently has the advantage in computer network operations. While this situation will shift in time more in favor of the network defender, planning network defense in today's world is starting from a disadvantage relative to the attacker. Some advances have been made by network defenders, making it more difficult for nonstate hackers to do as much damage as they previously did. However, the more advanced type of hacker, especially state-sponsored hackers, has the time and resources to prepare actions of great damage. To address this problem, we presented seven principles of network defense:

1. Prepare different forms of contingency plans in advance.
2. When responding to a threat, do less harm than good to your own network.
3. Plan and execute flexibility in response.
4. Have a fallback plan for how to handle surprises.
5. Plan and execute flexibility in visibility.
6. Provide defense in depth.
7. Identify and block attacker intent.

References

[1] Churchill, Sir Winston, "Sir Winston Churchill quotes," http://www.military-quotes.com/, April 2005.

[2] Tzu, Sun, "Sun Tzu quotes," http://www.military-quotes.com/, April 2005.

[3] Smith, Michael, "'Secret Weapon' Crafted at Rhode Island Mill Helped Keep Germans Off Balance," *Stars and Stripes*, June 6, 2001, http://ww2.pstripes.osd.mil/01/jun01/ed060601j.html.

[4] "World War II," Global Security Web site, http://www/globalsecurity.org/military/ops/world_war_2.htm.

[5] Arkin, William M., "The Gulf War: The Secret History," The Memory Hole Web site, http://www.thememoryhole.org/war/gulf-secret02.htm.

[6] Rouse, Ed, "The Gulf War," *Psywarrior.com*, http://www.psywarrior.com/gulfwar.html, February 2005.

[7] Gill, Cindy, Journalist 1st Class, "OPSEC Is Everyone's Responsibility," U.S. Navy Press Release, undated.

[8] Wernick, Robert, *Blitzkrieg: World War II Series*, Alexandria, VA: Time-Life Books, 1977.

[9] Elias, Ann, "The Organization of Camouflage in Australia in the Second World War," *Journal of the Australian War Memorial*, http://www.awm.gov.au/journal/j38/camouflage.htm#_ednref53.

[10] "Stormtrooper," http://en.wikipedia.org/wiki/Stormtrooper, May 2006.

[11] Acohido, Byron, and Jon Swartz, "Unprotected PCs Can Be Hijacked in Minutes," November 29, 2004, http://usatoday.com/money/industries/technology/2004-11-29-honeypot_x.htm.

[12] "DEFCON DEFense CONdition," http://www.fas.org/nuke/guide/usa/c3i/defcon.htm, July 2005.

[13] Uttenweiler, Bill, "Cyber Attacks Spur Approval of INFOCON Structure," Vandenberg, CA, March 1998, http://members.impulse.net/~sate/infocon.html.

[14] *Information Security: Challenges to Improving DoD's Incident Response Capabilities*, Government Accounting Office, Report # GAO-01-341, March 2001, http://www.globalsecurity.org/military/library/report/gao/gao-01-341.htm.

[15] *FBO Daily*, FBO #0949, July 2, 2004, http://www.fbodaily.com/archive/2004/07-July/02-Jul-2004/FBO-00613197.htm.

[16] Allen, Patrick, and Chris Demchak, "The Palestinian/Israeli Cyber War and Implications for the Future," *Military Review*, March–April 2003.

[17] Krebs, Brian, "FBI Arrests Hacker in Planned New Year's Eve Attack," *Newsbytes*, Washington, D.C., January 12, 2001, 4:54 PM CST; reported in Inforwar.com on January 15, 2001, http://www.infowar.com/hacker/01/hack_011501b_j.shtml.

[18] Krebs, Brian, "Feds Warn of Concerted Hacker Attacks on New Year's Eve," originally reported in *Newsbytes*, Washington D.C., December 29, 2000; reported in Infowar.com, December 29, 2000, http://www.infowar.com/hacker/00/hack_122900a_j.shtml.

[19] Lipowicz, Alice, "Report: Punish Poor Information Security Setups," *Washington Technology*, November 8, 2005, http://www.washingtontechnology.com/news/1_1/daily_news/27391-1.html.

[20] "Request for Information for Host-Based Security Manager Support," May 31, 2005, http://www.eps.gov/spg/DISA/D4AD/DITCO/RFI326/SynopsisR.html.

[21] Definition of Honey Pot, April 20, 2005, http://searchsecurity.techtarget.com/sDefinition/0,,sid14_gci551721,00.html.

8

Monitoring and Replanning During Execution

No plan survives contact with the enemy.
—Field Marshal Helmuth von Moltke

Chapter 3 described how executing a plan with feedback has a greater probability of success than executing a plan without feedback. This is true for all plans, and especially true for IO plans since the cause-and-effect chain can be very long from action to ultimate desired effect.

Section 8.1 presents the need for monitoring the progress of the plan to determine whether events are unfolding according to the plan. This section also raises the need for automated monitoring of events in the Information Age, so that the status information essential to success is readily available to the decision-makers. Speed of decision-making is essential, especially in many aspects of IO.

Due to the need for speed, planners traditionally develop contingency plans, often embedded within the main plan. Section 8.2 presents branches and sequels as mini-contingency plans and control measures within the overall plan. The conditions for the triggering of these branch and sequel decisions must be monitored to determine whether the conditions are being met, and whether additional effort is required to shape the conditions for the preferred branch or timing of a sequel.

Sections 8.3, 8.4, and 8.5 present execution monitoring concepts for attack plans, influence plans, and defense plans, respectively. Due to differences in the nature of attack, influence, and defense plans, the monitoring and replanning of each have some unique considerations.

When an operation deviates substantially from its plan, and the built-in branches and sequel contingencies cannot compensate for the deviation, it is time to replan. Section 8.6 addresses replanning. Replanning can range all the way from "back to the drawing board" to "modify this section of the plan." The risk of replanning is that it often creates a pause in the operation that allows the enemy to either take the initiative or better respond to conditions where we previously had the initiative.

Section 8.7 discusses the role of feedback from plan execution back to the planning process. All planning involves the use of planning factors to help define the scope and timing of different plan elements. Due to increased complexity in the Information Age, there is an increased need to rely on validated planning factors since there is rarely an expert in every field always available to support every element of the plan. However, the increased reliance on planning factors presents the need for a feedback process, whereby the planning factors are continuously refined based on the differences between observed versus predicted outcomes. Without this feedback process, plans will often predict unachievable outcomes, and commander and troop confidence in the plans will suffer.

8.1 Monitoring Plan Progress

Von Moltke's leading quote recognized that no plan will unfold exactly according to expectations. Enemy actions, acts of nature, and friendly mistakes all conspire to cause a plan to require mid-course corrections [1]: "There is no such thing as a perfect plan." The best plan is one that can proceed despite encounters with the unexpected. Therefore, the "best" plans are those that are robust and sufficiently flexible to address unforeseen circumstances. To quote Sir Basil Liddel-Hart [2]:

> In any problem where an opposing force exists and cannot be regulated, one must foresee and provide for alternative courses. Adaptability is the law which governs survival in war as in life .... To be practical, any plan must take account of the enemy's power to frustrate it; the best chance of overcoming such obstruction is to have a plan that can be easily varied to fit the circumstances met.

Since actual outcomes are likely to deviate from the plan, the plan must include a description of the means by which one intends to monitor the progress of events and determine in real time whether the plan is succeeding, requires modification, or is failing. In other words, one must plan for the monitoring of the plan. The more precise and quantitative the measures, the easier and quicker the comparison, and therefore the easier it will be for the warfighter to determine how closely (or not) the plan is matching the unfolding events.

8.1.1 Planning for Monitoring

The philosophy of execution monitoring is that the probability of success with feedback is greater than or equal to the probability of success without feedback. If you can respond to enemy or unplanned actions or effects, you have a better chance of achieving the objective. The first step toward increasing the probability of success is to define how and when one will measure the success of each element of the plan. These measures will inform the warfighter whether events are, or are not, unfolding according to plan.

Plan monitoring elements need to include measures of the timing of actions, the timing of effects, and when and how one plans to measure these effects. While the cause-and-effect network describes the purpose and relationship of various objectives, actions, and effects, it does not include a temporal representation. The most commonly used temporal representation in military planning is the *synchronization matrix*. The traditional "synch" matrix shows when given tasks are undertaken on a timeline (such as a Gantt chart) so that they can be deconflicted with other tasks. Moreover, dependent tasks can be displayed over time to ensure that each task is taken in the proper sequence. The synch matrix has been used successfully for over a decade. Originally developed in PowerPoint by V Corps in Europe, various electronic representations of the synch matrix have emerged over time.

A recent enhancement to the traditional synch matrix is the *Enhanced Synch Matrix*, developed by ManTech Aegis Research Corporation. The Enhanced Synch Matrix includes not only the start and duration of each friendly task, but also displays two additional attributes associated with each task. The first is the start time and duration of the *effect*, while the second is the planned *observation points*. We will describe the synchronization of the effects first.

In effects-based operations in general, and in IO in particular, there may be a significant lag time between the start and finish of a friendly action and the start and finish of its intended effect. For example, if leaflets are dropped, the end of the leaflet drop occurs well before the first enemy soldier deserts or surrenders. In other influence operations, the message sent to a group may have a significant delayed reaction before the desired effect is achieved. By explicitly distinguishing between the time frame of the action and the time frame of the effect, better *effect synchronization* can be achieved. Much as time on target has a greater effect in artillery bombardment, synchronized effects have a greater impact on the intended target than unsynchronized effects.

The second new feature in the Enhanced Synch Matrix is the identification of desired observation points. Since every plan needs to measure progress, the measurement needs to be observed. In many cases, the action taken can be observed and confirmed by the personnel performing the action. However, in the field of IO, automated actions may still require independent confirmation of the successful completion of the action, as this might not be directly observable by a human. Moreover, if the start of the effect is separated in time from the completion of the action, measuring the achievement of the effects is often more complicated and requires separate observations. These observations of the effect must be planned for, and the time sensitivity and time tradeoffs need to be tracked. The three features of the Enhanced Synchronization Matrix are shown in Figure 8.1.

Planning for observations does not mean that the operational planner is tasking the reconnaissance assets. In many cases, the necessary intelligence assets are owned by another organization. Having the planner plot the desired observation points on the synch matrix helps show and explain the purpose and the tradeoffs for this particular observation request or request for information (RFI). By keeping the request in the context of the rest of the plan, the intelligence asset planners can see the tradeoffs between balancing the requested timing of an

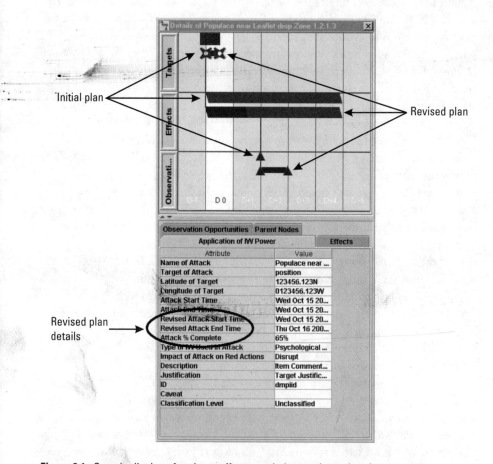

Figure 8.1 Sample display of actions, effects, and observation points for one action.

observation against the constraints of scarce sensor assets. Because Enhanced Synch Matrix shows the duration of the anticipated effect, and the requested observation has a rationale, the intelligence asset planner can see whether changes in the time of the observation will still allow the required information to be gathered.

8.1.2 Monitoring Plan Execution

Once the plan is completed, including its synchronization of effects and observations, the monitoring of the execution of the plan can begin. This means that actual status data needs to be inserted into some command-and-control system to be able to compare the planned results versus actual results. Traditionally, the situation map, covered by plastic and grease-pencil marks, was the focal point for situation awareness and determining whether events were unfolding according to plan. Even though electronic maps and displays are now being used, almost all plan monitoring and status updates are still performed manually. Moreover, the situational awareness data is still collected primarily via voice radio [3]. Since high-bandwidth data communications are not yet prevalent and sufficiently mobile on the battlefield, voice communication continues to be the most reliable source of plan status information.

A similar manual update process occurs in IO plan execution. The success or failure of a given action is placed manually into the computer used to monitor the situation. This leads to delays in understanding the current situation, and therefore in the comparison of planned versus actual outcomes. This, in turn, leads to delays in knowing whether the current plan is working, needs modification, or is failing.

In the future, we predict that the ability to collect automated plan status information will continue to increase. Blue Force Tracking (BFT) technology is becoming more widespread and accepted. During Operation Iraqi Freedom, better Blue Force situation awareness avoided fratricide and increased the rate of engagements against the enemy [3]. Moreover, DARPA's Command Post of the Future (CPoF) is a new tool in use in Iraq that allows for selected data items, such as Blue Force Tracking, to be received directly into the tool for more timely situation awareness. Once the real-time force and status tracking is commonly fed directly to plan monitoring tools, we will start to see the ability to compare expected versus actual outcomes in real time or near real time.

In the meantime, some combination of automated and manually input data will be displayed on computer screens to provide near-real-time situation awareness and therefore will better support the ability to compare planned versus actual outcomes. An Execution Monitoring Tool will be required to perform this function, as it needs to display not only the plan information, but also the observed outcomes. An example of the features of an Execution Monitoring Tool is shown in Figure 8.2.

In Figure 8.2, the following information is available for each task and its associated desired effect:

- Planned action start and end time;
- Actual action start and end time;
- Planned effect start and end time;
- Actual effect start and end time;
- Planned observation times and measurables;
- Actual observation times and measurables.

Additional features are also possible, such as percent of task completed, alternate observation times and measures, and revised planned start and end times. The key is to populate the display with enough information to allow the warfighter to comprehend planned versus actual status at a glance, without overloading the warfighter with too much information.

Note that if one task is not going according to schedule, it may or may not delay "downstream" tasks. This is a function of how dependent the downstream tasks are on the delayed task. A 2-hour delay in the first task may cause a 2-hour

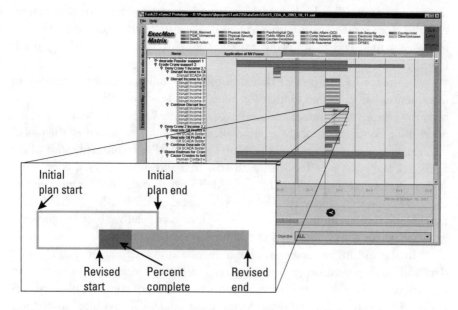

Figure 8.2 Sample execution monitoring display of planned versus actual events.

delay in a later task, or a 4-hour delay, or only a half-hour delay, or no delay at all. It all depends upon how the delayed achievement of the desired end state of one task affects the start time of later tasks. There is no easy, clear way to define this relationship in an automated fashion. As a result, each delay must be considered manually with human judgment to determine the effect on the downstream tasks.[1]

High-speed accurate information collection and exchange are essential in most IO attack and defense operations, and in selected influence operations. In the technical realm, computer network attack or exploitation opportunities can be fleeting, while in computer network defense, a slow response is ineffective against a skilled attacker. In the influence realm, being the first with a message gains the initiative in the idea battlespace, forcing everyone else to respond to that message. The ability to provide automated plan status will revolutionize the ability to operate in the information domain.

In addition to displays on the comparison of planned versus actual actions, effects, and observation points, the Execution Monitoring Tool needs to display upcoming decision points, whether their conditions for making each decision have been met, and whether additional action is required to change the conditions before the decision is made. These points are addressed in Section 8.2, since decision points are usually associated with plan branches and sequels.

8.2 Monitoring Branches and Sequels

Branches and sequels are mini-contingency plans embedded within the plan being executed. Branches and sequels allow for the smooth continuation of operations without having to stop, evaluate the situation, and then figure out what to do next [4]: "Every plan of campaign ought to have several branches and to have been so well thought out that one or other of the said branches cannot fail of success." During planning, the conditions and options available have already been evaluated, and taking a particular branch provides a seamless continuation of operations. It is like coming to a fork in the road and continuing to drive down one path without having to stop and read the map.

While branches and sequels were introduced in Chapter 5, we present more details about them here, including how they are used during plan execution.

---

1. One could try and preplan what the effects of delays will be, which would then make the implementation of ripple effects straightforward and automated. However, this requires the planner to manually make that evaluation during planning. In either case, whether done during planning or during plan execution, the process of determining the downstream effects of delays on earlier tasks is currently a manual process.

8.2.1 Sequels

The Department of Defense (DoD) defines a sequel as [5]: "A major operation that follows the current major operation. Plans for a sequel are based on the possible outcomes (success, stalemate, or defeat) associated with the current operation."

For execution monitoring purposes, we have modified this definition slightly to be more comprehensive, and therefore more useful to all echelons. For example, Joint doctrine currently defines a sequel as a major operation that follows the current major operation. There is no reason, however, why the planning and execution process for sequels should not be applicable at all echelons of planning. There are a number of situations in IO where a sequel in a plan is useful, even if it is for a minor operation.

For example, a sequel could be defined to exploit a certain enemy information source once undetected access has been confirmed. The exploitation step is a sequel to successful access, but the conditions for exploitation may include confirming that the access was undetected by the enemy before exploitation is allowed to take place. Therefore, the exploitation step in this case qualifies as a sequel even though it is not a major operation.

In our expanded definition, sequels are defined as *actions to be taken only after specified conditions have been met*. The only question is *when* the sequel will be initiated. Whether a specific sequel is activated is a function of whether its triggering conditions have been met. There are no branches to sequels. The action is either taken or not taken at any given time.

Sequels are defined during planning, and are triggered during the execution phase when the specified conditions have been met. The conditions may also be overridden by a commander due to changes in the circumstances. For example, during Operation Desert Storm, General Schwarzkopf launched the main thrust of the left hook many hours earlier than originally planned because the Marine right wing was advancing so quickly. To ensure that the Iraqis were caught in the pocket, the triggering conditions were considered by the commander to have been met earlier than previously planned [6]. Note that the other triggering conditions (location of enemy forces relative to our own) remained the same, and only the time condition was overridden in this example.

Our definition of a sequel thus includes the DoD definition and expands it further to explicitly include sequel conditions that currently occur implicitly within all echelons of the plan, but that may already be called something else in practice. For example, the Army uses *phase lines* to control the coherent advance of forces. If one thrust reaches its phase line before its flank elements do, then it waits for the flanks to catch up. Note that the conditions for which to take a subsequent action will be explicitly specified as a sequel, but an Army officer would interpret it as being standard operating procedures for phase line operations. However, in order to place such standard operating procedures in electronic format and allow for automated monitoring of plan status, we are taking control measures that are currently separate, such as phase lines and sequels, and combining them into a single definition of sequels that can be clearly defined, monitored, and executed when the predefined conditions are met.

Sequels are explicit elements of a cause-and-effect network. The sequel conditions and triggering decision will apply to all subordinate tasks and actions downstream in the CAEN. This list of conditions may include any combination of time, location, and friendly or enemy status. Due to the wide range of possible conditions and combinations, the format for the conditions will be fairly free-text, with specific key words such as Booleans and if-then statements. Since planners and warfighters normally define conditional statements of "When this occurs, then do that," this approach does not introduce a unique format to DoD.

Military Standard 25-25 does not appear to have any symbology for sequels. We selected a symbol for sequels as described in Chapter 9, but this candidate symbol needs to be officially approved by the proponent of Military Standard 25-25. It does not matter which symbol is used for sequel, as long as the symbol connects to the conditions required to trigger the sequel and the current status of those conditions.

8.2.2 Branches

The Department of Defense (DoD) defines a branch as [5]: "The contingency options built into the basic plan. A branch is used for changing the mission, orientation, or direction of movement of a force to aid success of the operation based on anticipated events, opportunities, or disruptions caused by enemy actions and reactions."

Branches are *alternative actions taken as a result of specified conditions being achieved.* Unlike sequels, branches have two or more options, of which only one will be selected at a decision point. For example, the force may go left or it may go right once the decision point is reached and the conditions for selecting one of the options have been met. As with sequels, a branch has a set of predefined conditions that must be met before one of the alternative branches is selected. Note that "wait until one of these conditions is met" is similar to a sequel, but the fact that branches have alternative actions that may be taken depending on the conditions makes them different from sequels. Like sequels, the commander may override the conditions specified in the branch definition and cause the triggering of the event. Note that our definition of a branch is functionally the same as the DoD definition, but focused on electronic implementation of a branch. What we have added are methods of quantifying the condition definitions for more automated status tracking and displaying the decision points and their dependencies.

Branches often have only two options, but may have more than two. As a result, many possible options may emerge from a single branch decision point. For sake of manageability and simplicity, it is counterproductive to have too many branches in a plan, or a large number of options for each branch. Note that for any N-option branch, there are N+1 states to consider—each of the N options and one for the pretrigger state where the decision is to "not decide yet."

The symbol representing a branch in a CAEN is a distinct icon. Multiple options can emerge the branch point in the plan, but only one option is actually implemented based on the predefined set of conditions. As with sequels, the branch symbol needs to link and provide access to the conditions that define the branch triggers, and the current status of those conditions. Figure 8.3 shows a sample branch point and sequel link.

There is not yet a military symbol for a branch with different paths, because Military Standard 25-25 does not yet include cause-and-effect networks. Military Standard 25-25 does include the symbol for a decision point, which is usually indicated by a star on the synchronization matrix. We include decision points on our synchronization matrix, as shown in Section 8.3. In addition, we include a display of the dependence of decision points downstream from each other, if they are indeed related.

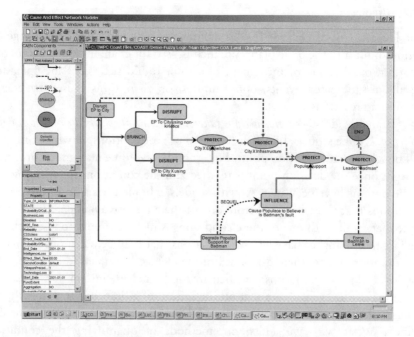

Figure 8.3 Sample branch point and sequel link in a cause-and-effect network.

8.2.3 Monitoring Decision Points

Besides the branch symbol on the CAEN, branch symbology is required in the Execution Monitoring Tool. The decision points (both branches and sequels) may have dependencies with respect to each other. For example, taking the north fork will preclude subsequent decision points associated with the south fork, but will also place increased emphasis on monitoring the conditions for the subsequent decision points that occur after the north fork is taken. Figure 8.4 shows an example of a decision point dependency display in the Execution Monitoring Tool.

Another useful display is the *status of the conditions* of each branch or sequel. Since the decision of when to launch a sequel or choose which branch to take is a function of the conditions at that time, monitoring the status of the conditions before the decision is an important capability. For example, if the preferred branch requires a condition for an enemy force to be neutralized, then taking steps to ensure that enemy force is neutralized before the decision point is reached will help ensure that the preferred branch is taken. For an historical example of a sequel, General Schwarzkopf specified that the Iraqi Republican Guard was reduced by air attacks to some specified amount before the ground war was to be launched [6]. Figure 8.5 shows a sample display of the status of conditions for a branch decision point.

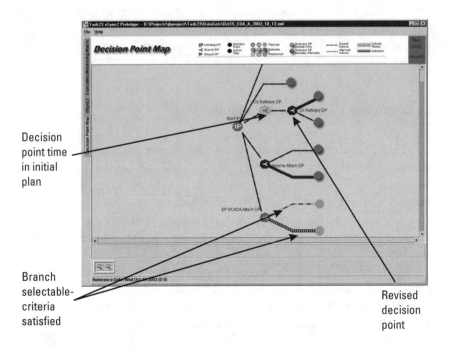

Figure 8.4 Decision point dependency map.

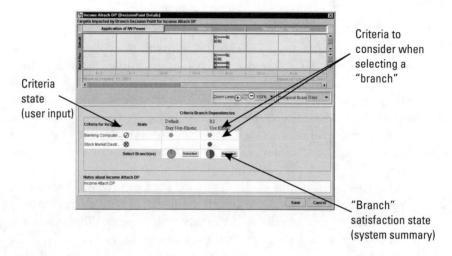

Figure 8.5 Sample display of status conditions for a branch decision point.

Overall, execution monitoring includes not only the monitoring of actual versus planned outcomes, but also includes the monitoring of key decision points and the status of conditions prior to reaching each decision point. Moreover, planners need to plan how to monitor the success of the plan, including the measures of success for actions, their associated effects, and the observation points required to confirm the status of the actions and effects. Lastly, the future ability to automatically monitor current status and compare it to the plan will help increase the speed of both situation awareness and key decision making.

8.3 Execution Monitoring for Attack

Offensive information operations, like all offensive action, carry the advantage of the initiative. The attacker chooses when, where, and how to strike, while the defender responds. As long as the attacker retains the initiative, the defender is constantly reacting. Unless the defender has some plan of his own to wrest the initiative from the attacker, or a plan to exploit the situation when the attacker reaches his culmination point, the attacker's plan will likely succeed or at worst face a draw.

Therefore, much of the attacker's plan is based around retaining and exploiting the initiative. Contingency plans, including branches and sequels within the plan, are a good way to help maintain the initiative for the attacker. For example, if the attacker is sure that the defender has only two ways to respond to the attack, then the attacker can implement the branch that best exploits whichever decision the defender makes. As long as the attacker can foresee feasible enemy actions, the attacker can sustain the initiative regardless of enemy efforts to react.[2]

An attacker's plan should include sequels to coordinate different actions and keep them synchronized, so that the attack components do not get out of step. If they were to get out of step, then the objective could be missed. For example, pushing the Iraqi forces out of the pocket in Kuwait would have missed the objective of trapping the Iraqi forces. In a similar manner, tracking the conditions on the branches helps ensure that whichever way the enemy, or nature, reacts, the initiative can be retained and the objective achieved. As mentioned previously, branches and sequels help keep the attacker from having to stop and read the map, as such delays could cause the loss of the initiative.

In information operations, the attacker's monitoring of the progress of the plan is a function of the timescale of the operation. In computer network attack (CNA), the time frames of interest can be very short, but can also include longer-term elements. For example, the time to identify the target, gain access, insert malicious code, further refine target identification, and exploit the target can be very short, especially if the probability of detection or attribution does not matter. Brute force attacks can successfully gain access for a short period of time, but the longer-term value of the penetration may be lost simply because the defender is aware of that penetration. Conversely, a CNA plan can be designed to take a long time. If the defender has substantial computer network defenses, the attacker may have to take each step very slowly in order to avoid detection. Moreover, the attacker can install leave-behinds to help facilitate later reentry and exploitation, which is a common hacker technique.

In either case, monitoring the progress of each task in the plan is important in determining whether the plan is succeeding as is or needs modification. In the fast attack scenario, it may be irrelevant to determine whether the enemy detected the penetration or exploitation activity. In the slow attack scenario, the attacker may include a series of observation points to determine whether the enemy has detected each of the attacker's completed steps. Of course, the attacker can never be sure that the steps were undetected, but just as in traditional military operations, the attacker cannot self-incapacitate for fear of being observed. Some risk and audacity are required in any attack, including CNA.

For electronic warfare, similar types of attack plans are often available. It does not matter whether brute-force EW attacks, like barrage jamming, are detected. Stealth is not the primary purpose of such an electronic attack. In other cases, EW needs to be comparatively stealthy, much like a CNA slow

---

2. That is why a spoiling attack by the defender can be so confusing to the attacker. The attacker is not sure if the defender is better organized than expected, or the force disposition is not as expected. A vigorous counterattack by the defender, even if unsuccessful, can significantly disrupt the confidence of the attacker.

attack. In either case, the plan elements and their observation points need methods of confirming whether the objective of a given task and its desired effects have been achieved.

Another form of mini-contingency planning is to specify probable Red reactions to the Blue action, and have preplanned Blue counters to these reactions. For example, if bombs have fallen near an enemy cultural site, be prepared for the enemy influence apparatus to try and exploit the event. For example, during Operation Desert Storm, the Iraqis damaged their own mosque to in an effort to claim that it was Allied bomb damage. The description of how this was discovered is described well by Mr. W. Hays Parks, chief, International Law Branch, Office of the Judge Advocate General of the Army [7]. He is referring to a series of images culminating in Figure 8.6.

> We were very careful in watching where we carried out our air strikes during the Gulf War and watching for collateral damage as well. This is the Al Basrah Mosque. It attracted our attention early in the air campaign because there was a bomb crater adjacent to it. In tracing its origin, it was determined that it occurred because of an errant bomb off of a Navy strike aircraft. This was not the target. The target was a military objective nearby. We wanted to make sure there was no more collateral damage done. We kept looking, and suddenly the next day, the mosque was falling apart. There had been no additional air strikes, but it was collapsing. Three days

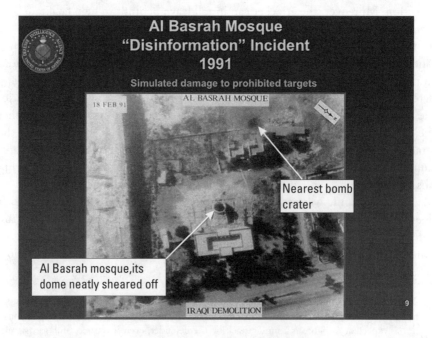

Figure 8.6 Example of faked damaged against a mosque, 1991 [8].

later, there was even more damage. Finally, the place is completely obliterated. Now what clearly was happening was the mosque was being torn down by the Iraqis so that Saddam Hussein could use this as a way of trying to split the Coalition because of the number of Moslem nations that were involved in it.

The Iraqi accusation that the mosque was damaged by Coalition bombs was disproven by having immediate and continuous battle damage assessment that showed the crater and the undamaged mosque in the first photo and later photos that showed the damaged mosque and no new craters. Planning ahead for effective battle damage assessment is important not only in terms of the decision whether to restrike, but also in collecting the evidence to counter possible enemy propaganda efforts.

Note that the mini-contingency plans for Red reactions and appropriate Blue counterreactions cross multiple IO domains. The preceding examples describe how to protect against enemy reactions in the idea battlespace to actions taken in the physical battlespace. The offensive IO plan needs to include contingencies across all three IO categories—attack, influence, and defense.

8.4 Execution Monitoring for Influence

The influence elements of the plan have their own unique features in terms of execution monitoring. First, as mentioned earlier, influence must be considered in all plans (not just IO plans) due to the fact that the battle for the minds never ends. Any physical action, offensive IO action, or even defensive IO action is subject to exploitation by the enemy for purposes of influencing minds. To this end, every plan needs to monitor for changes in the idea battlespace, to plan ahead for probable Red reactions, and have Blue counterreactions prepared to quickly respond to those actions. As mentioned in Chapter 6, the idea battlespace is constantly receiving new messages and sometimes receives new groups into the arena. It is critical to any plan to constantly monitor the idea battlespace to predict which messages the enemy will create and how to best counter them.

As with all other arenas of conflict, gaining and retaining the initiative in the idea battlespace is essential to success. In the idea battlespace, being the first with a message gains the initiative, forcing everyone else to respond to that message. Getting and maintaining attention in the face of competing messages and boredom with the original message are also important. These are some of the key measurables that need to be defined in the idea battlespace portion of the overall plan and monitored for success or failure.

Branches and sequels are as important to the success of plans in the idea battlespace as with any other battlespace. If this type of message appears, do that. If this group is brought into the arena, send that message. Speed and accuracy of response are critical to success. Gaining and maintaining the initiative are also essential to success, while delays and muddled responses are a boon for the enemy. Other useful mini-contingency plans include preplanned responses to predicted Red reactions to our messages. With enough foresight and rapid response capabilities, one could choreograph maneuvers and conflicts in the idea battlespace much the same way as the U.S. military currently dominates the physical battlefield. However, much needs to be done in developing the doctrine and capabilities necessary to compete effectively in the idea battlespace, as described in Chapter 11.

The idea battlespace is changing quickly in today's world. The global reach of communications and media broadcasts means that messages reach more people more quickly than ever before. Therefore, it is essential to constantly monitor for changes in the idea battlespace, and especially to identify changes that were not predicted in the plan. In some cases, the change is an opportunity to exploit an enemy error, but such errors must be exploited quickly before attention on the subject degrades.

Lastly, planning in the idea battlespace is probably most dependent on identifying the correct measures of cause (MOCs). The measures of performance (MOPs) are well understood for delivery mechanisms, such as dropping leaflets or broadcasting on frequencies. The measures of effectiveness (MOEs) for achieving desired effects in terms of public opinion are also fairly well understood. What is often lacking is connecting the dots—defining how the action will translate into the desired effect. This causal linkage requires its own measure so that the assumed relationship between action and effect can be explicitly evaluated. It is during execution monitoring that any potential flaws in the assumed causal linkage will be detected. The sooner they are detected and confirmed, the sooner corrective action can be taken.

As an example, let us examine the hotel bombings by al-Qaeda in Jordan on November 9, 2005. When the terrorists found a wedding party at the targeted hotel, they did not care that the attendees were Palestinians and Jordanians, nor that women and children were present [9]. They simply wanted to inflict the maximum number of casualties possible using explosive-propelled ball bearings [10]. The outcry against al-Qaeda by the Jordanians and even ranking Palestinians was enormous [11]. In response, al-Qaeda tried to backtrack and claim that Jordanians and Palestinians were not the target [12]. However, the cold-blooded murder of innocent men, women, and children spoke clearly of their motives in contrast to their postevent denials [13].

This example showed al-Qaeda's assumed causal relationship that maximum casualties would be best for al-Qaeda, get the most coverage in the media, put pressure on the Jordanian government, and possibly force Western interests out of Jordan. Moreover, vocal critics of the West would frequently chime in after any such incident and blame the West and its allies for being the root cause of the problem that led to the bombings, further reinforcing al-Qaeda prestige. None of these expected outcomes occurred. By intentionally targeting a wedding party, al-Qaeda turned public opinion in Jordan, Palestine, and around the world against them. The Western media focused on the callousness of the attack instead of the faults of the West. The Jordanian government was strengthened, not weakened. And while a short-term financial downturn hit Western interests in Jordan, there were, to date, no longer-term repercussions. Lastly, al-Qaeda's feeble attempts to claim that killing Middle Eastern civilians was not the purpose were clearly shown to be false based upon their instructions to their terrorist agents. The assumed causal relationship between action and effect backfired against al-Qaeda in this incident, and they were ill prepared to compensate for the disaster in the idea battlespace.

Overall, monitoring for influence operations includes monitoring the idea battlespace for ground truth, comparing this to the measurables in the plan to the observed results, and determining whether changes are in order. Moreover, due to the fluidity of the idea battlespace, constant monitoring of the arena is essential to ensure that key new messages are detected, and that the entry of new groups is detected as well. Messages and group entries into the idea battlespace must be predicted, not just detected, so that the reactions and the shaping future events are already underway by the time the enemy messages are launched or the new group arrives. Contingency plans, including branches, sequels, and predicted Red reactions and planned Blue counterreactions, are all elements to ensuring rapid and effective response and to retain the initiative in the idea battlespace (while being first with a message helps gain the initiative). But all of this takes constant monitoring and rapid response capabilities—capabilities the United States has not yet mastered. (See Chapter 11 for more discussion of this last point.)

8.5 Execution Monitoring for Defense

There are four core categories of defensive IO: military deception, operations security, computer network defense, and electronic protection (defensive electronic warfare). Each of these categories has unique execution monitoring considerations, as described next. Monitoring MILDEC and OPSEC plans is described in Section 8.5.1, while monitoring CND plans is described in Section 8.5.2. Elements of electronic protection are included in the OPSEC section.

8.5.1 Monitoring Military Deception and OPSEC Plans

Chapter 7 described how to plan military deception (MILDEC) operations and operations security (OPSEC) as adjunct plans supporting the main plan. The planner defines a set of signatures he wants the enemy to see in the MILDEC plan and a set of signatures he does not want the enemy to see in the OPSEC plan. The planner also indicates in the plan how he intends to measure whether the enemy has seen what he was intended to see and has not seen what he was not intended to see.

During plan execution, all of these signatures and indicators of enemy detection need to be constantly monitored (or regularly monitored within a reasonable cycle time). For example, a Blue intelligence cell can monitor the signatures to see what the enemy might have detected. In the case of military deception plans, this is a necessary but not sufficient condition to determine whether the enemy actually detected the bait signatures. Table 8.1 provides a sample deception task of "Our focus is to destroy enemy units." Such a task would likely generate four signatures, which are listed in the first column. The second column indicates the types of enemy sensors that could detect each type of signature. The third column provides the current measure of whether the enemy sensor actually detected, or could have detected, the signature that we wanted the enemy to see.

In the case of OPSEC, detecting friendly signatures that should not be released means that the enemy might have detected them, but does not mean that they actually did. (See Table 8.2.) Other intelligence sources are required to make that determination. However, friendly monitoring of friendly signatures, both as part of the deception plan and as part of confirming the success of the OPSEC plan, is essential to monitoring the success of these plans.

Like Table 8.1, Table 8.2 tracks the signatures we think the enemy might have detected, but these are the signatures we did not want them to detect. In this example, Blue wants to capture occupied oilfields intact. The four supporting actions will have signatures we want to hide from the enemy. The first column lists the signatures for each action, while the second column indicates the types of enemy sensors that might be able to detect those signatures. The third column describes the current status of what we believe the enemy has or has not detected of these four signatures.

As mentioned in Chapter 7, the military deception plan does not have to be perfect—it just has to be better than any mistakes leaking through the OPSEC plan. Although each plan has its own signatures to monitor and measurables to determine, the ultimate success or failure of the military deception and OPSEC plans is considered as a whole. Either the enemy did not detect the actual plan in time to respond adequately, or the enemy did, thereby causing the actual plan's intent to be compromised to the enemy.

Table 8.1
Sample Signatures and Indicators of Enemy Detection for MilDec

Our Focus Is to Destroy Enemy Units	Red Sensors That Could Detect Signature	Current Measure of Success (Detected?)
Signature 1: Possible SOF mission insertion observed	Red sensor type 1 (e.g., radar) Red sensor type 2 (e.g., aerial detection)	SIGINT confirms enemy detected it No enemy aircraft flying—no detection
Signature 2: Two enemy HQs hit by laser guided bombs	Red monitoring of media broadcasts	Red known to receive media broadcasts
Signature 3: Blue maneuver units appear to be avoiding contact with Red units	Red sensor type 2 (e.g., aerial detection) Red sensor type 3 (e.g., spies or collaborators)	No enemy aircraft flying—no detection HUMINT double agent confirms enemy detected it
Signature 4: Blue units appear to be waiting	Red sensor type 4 (Red situation reports) Red monitoring of media broadcasts	No confirmation yet Red known to receive media broadcasts

Table 8.2
Sample Signatures and Indicators of Enemy Detection for OPSEC

Capture Oil Fields Intact	Red Sensors to Avoid or Block	Current Measure of Success (Undetected?)
Signature 1: Possible SOF mission insertion unobserved	Red sensor type 1 (e.g., radar) Red sensor type 2 (e.g., aerial detection)	No SIGINT report of insertion event No enemy aircraft flying—no detection
Signature 2: Demolitions neutralized	Red guards and demolitions experts	Demolitions not replaced
Signature 3: Maneuver to target unobserved	Red sensor type 2 (e.g., aerial detection) Red sensor type 3 (e.g., Red scouts)	No enemy aircraft flying—no detection No enemy scouts detected in area
Signature 4: Presence at oil fields unobserved for at least 4 hours	Red sensor type 3 (e.g., Red scouts) Red situation reports	Enemy scouts detected at hour 6 Blue jamming Red communications

The good news is that during plan execution, the degree to which the military deception plan is working, or the OPSEC plan is not working, can be measured mid-course, so that corrective action can be taken. Like any other plan, these measurables help the warfighter determine whether the plan is working, and if not, in what specific ways. If there is an OPSEC leak, try to take corrective action to wind it back into the deception plan. If an element of the deception plan is exposed, have a contingency plan ready to explain it away.

As with any other type of planning, contingency plan elements should be embedded with the military deception and OPSEC plans. In contrast to anticipated Red reactions described previously, in most cases these are preplanned Blue reactions to foreseeable problems in the military deception and OPSEC plans.

In addition, military deception plans and OPSEC plans may have sequels, such as ensuring that one part of the plan is implemented before the next begins. Moreover, there may be *interconnections* between the military deception and OPSEC plans that require one part of the OPSEC plan to be implemented before another part of the military deception plan is implemented, or vice versa.

Overall, the monitoring of the military deception and OPSEC adjunct plans to the main plan allows for mid-course corrections to provide for a greater probability of success. Moreover, efficient and effective monitoring of the two plans can be achieved by using shared resources for monitoring friendly signatures and determining which plan the enemy appears to be believing.

8.5.2 Monitoring Computer Network Defense

In computer network defense, the initiative has been ceded to the attacker. As mentioned in Chapter 7, performing a counterstrike in cyberspace can be risky, as the source of the attack may be disguised, and innocent parties may suffer the results of the counterattack. Therefore, most counteroffensive action in CND involves actions taken against intruders and malicious software within the defended network.[3]

Information control (INFOCON) was described in the previous chapter as a form of contingency planning for CND. At each INFOCON level, a set of acceptable actions are allowed to each user on the network. The lower INFOCON levels are more permissive, while the higher INFOCON levels are more restrictive. When enough conditions are observed to trigger an increase in INFOCON level, the new conditions are applied to the network. Like branches and sequels, INFOCON network defense monitors conditions and launches a predefined or contingency response to that specified set of conditions.

---

3. Note that we are not covering the investigation and arrest of hackers by civilian authorities, but focusing on defending military networks from opposing nation states and terror organizations.

In addition to INFOCON, computer network defenders can also include sequels, such as "when this happens, do that," or even branches. For example, in a honey pot, the defender may choose to continue to monitor the intruder if the damage is minimal, or the defender may choose to shut down the intruder if he is posing a particularly dangerous threat.

As in attack and influence, speed of decision and speed of action are important to CND. A slow response to a skilled attacker is not likely to be useful in stopping damage or compromise. One way to increase the speed of response is through predefined conditions and responses, as defined earlier. Another advantage of preplanned responses is that the effect the response will have on the friendly network is already known. Therefore, there is less concern that the response will do more damage to the network than the attacker could, because the response has already been analyzed and approved.

8.6 When to Replan—or Not Replan

To replan or not to replan—that is the question. Apologies to Shakespeare, but the decision of when to start replanning is an important and serious one to make. It is not a decision taken lightly. A great deal of time and resources has been invested in the current plan. Taking the time to replan risks losing the initiative to the enemy. Moreover, replanning when such action is not necessary consumes resources and can miss fleeting opportunities. To quote Clausewitz [14]:

> After we have thought out everything carefully in advance and have sought and found without prejudice the most plausible plan, we must not be ready to abandon it at the slightest provocation. Should this certainty be lacking, we must tell ourselves that nothing is accomplished in warfare without daring; that the nature of war certainly does not let us see at all times where we are going; that what is probable will always be probable though at the moment it may not seem so; and finally, that we cannot be readily ruined by a single error, if we have made reasonable preparations.

In some cases, a minor modification to a plan is all that is needed, and so full-scale replanning does not have to be undertaken. The cases of overriding one of the plan factors described previously in this chapter are examples of minor modifications that did not require replanning. However, there will be situations in which the operation is deviating substantially from its plan, and the built-in branches and sequel contingencies cannot compensate for the deviation. That is when it is time to replan.

Replanning may be limited to a portion of the plan, or to the whole plan. If only a portion of the plan needs to be replanned, then the initiative can be retained in the other areas of the plan. There may also be ways to compensate for delays in the areas requiring replanning. For example, if the termination of the current operations under the existing plan is farther in the future than the expected completion time of the revised plan, then the initiative might not be lost.

One aspect that helps in this decision process is having clear measurables for the plan, so that the warfighter can quickly and easily recognize when the plan has deviated too far to be recoverable. Defining measures as objectively as possible helps avoid as much subjective bias as possible. In the end, however, the decision to replan is so significant that it should remain a human decision. Some useful measures to consider include:

- If conditions are such that none of the condition sets necessary to select a key branch in the plan will occur in the foreseeable time frame. Since branches are decision points, then the decision to take any branch will never occur if the prerequisite conditions are no longer achievable.
- If conditions are such that a major preplanned sequel will never occur. As with the preceding example, the commander will either wait indefinitely for the conditions set to be fulfilled, or replan to address the new reality.
- If none of the key objectives has been achieved in the specified time frame, and there is no hope of achieving those objectives in a reasonable revised time frame.
- If the resources to achieve future planned tasks are no longer available, and simple on-the-fly resource reallocation is insufficient to resolve the shortages.
- If the situation has changed so dramatically that even achieving the plan's objectives is either infeasible or no longer relevant. For example, if the enemy caught you by surprise and now has the initiative, it is probably time to replan.

The decision to replan for IO follows these principles. Each type of IO, whether offensive, influence, or defensive, needs measurable interim points to determine whether events are unfolding according to the plan. On the offensive side, it may be that none of the Blue desired effects are being achieved against the targeted Red activities. On the influence side, it may be that none of the desired beliefs or actions is occurring according to schedule. On the defense side, it may be that there have been too many OPSEC leaks or the enemy is not believing the military deception plan. In each case, the key to the decision of whether to replan and why can and should be traced to the measurables of the plan, including specific interim objectives, branch and sequel conditions, resource status, and a new and unexpected situation.

8.7 Feedback Process to Refine Planning Factors

Planning involves the use of planning factors. How soon can I expect a certain result? How broad an effect will this option have? How long will the effect last? Due to the broad nature of IO, there is an increased reliance on validated planning factors to help planners plan. Since there is rarely an expert in every field available in the combatant commander's staff, planning factors and reachback support are the two methods used to provide the necessary expertise to the planners.

As mentioned in Chapter 3, feedback is an essential part of IO planning. In the preceding sections of this chapter, we discussed the use of feedback to support execution monitoring and preparing for key decision points. While the examples in Chapter 3 focused on feedback during a mission, there is another important aspect to feedback: improving the planning factors.

For plans to be effective, their expected outcomes must be realistic. If the planning factors are overly optimistic, increased risk will occur and problems will almost certainly arise. If the planning factors are too pessimistic, fleeting opportunities will be lost. To ensure that the best planning factors are available, a feedback process needs to be set up so that after an operation, planners can carefully analyze the differences between actual and expected results. When discrepancies arise, either the existing planning factors need to be modified, or some new qualification or categorization of when to use these planning factors needs to be applied.

In the first case (modifying planning factors), if the plan stated that under these terrain and weather conditions, the unopposed movement rate would be 24 kilometers per hour for a company-sized unit, and the actual rate was 20 kilometers per hour, and there appears to be no problem with unit training or execution, then the planning factor may need to be revised downward. However, if there is a specific reason that the slower rate occurred and that condition was unique and not widely applicable, then the planning factor may not need to be revised downward. At the same time, a track record of deviations from the planning factor needs to be evaluated over time to see if there is always some reason why the 24-kilometer-per-hour movement rate is not being achieved.

In the second case, there may be a need to add new qualifications or categories to the planning factors. For example, if the planning factor for the probability of access against an enemy network is 80%, and the actual results are 30%, then the reason needs to be carefully examined. It may be that the probability of gaining access against certain types of networks is 99% and the probability of gaining access against other kinds is 10%, whereupon the planning factors needs to be split into separate categories for each type of network. This increased level of specificity helps avoid the problem of never encountering the average situation that the first planning factor assumed.

Note that these evaluations are intended to occur after the conflict, or at least after the most demanding times in the conflict. The combatant commander staffs do not have time to collect and analyze planning factor data. This effort needs to be accomplished by an organization responsible for developing these factors. For example, the providers of the *Joint Munitions Effectiveness Manual* (JMEM) are likely candidates for refining the planning factors for kinetic options. In time, the group assigned to develop the *Information Operations JMEM* (IO-JMEM) may be able to provide similar support to refining IO planning factors.

After World War II, the U.S. Army published FM 101-10-1, *Staff Officers' Field Manual: Organization, Technical, and Logistical Planning Factors*. This field manual was the planner's bible for many decades. However, as the nature of war evolved in the Information Age, this manual was retired. New sources of planning factors need to be developed for all the traditional and IO options available to the modern commander and his staff. Moreover, there need to be a process in place and an organization assigned to monitor whether the existing planning factors are working or if they need revision. If they need revision, is the solution simply to change the value of the planning factor, or is it to redefine the conditions under which the planning factors are applicable? In either case, how well each planning factor performs will have a direct impact on the efficiency, effectiveness, and morale of the forces that use these planning factors. When good planning factors are used, planners and the troops they plan for will have confidence in the planning factors, and the missions planned with those factors.

Overall, the purpose of execution monitoring is to quickly determine whether actual outcomes are unfolding according to the plan, or if there are serious deviations between the two. In addition to monitoring the overall success of the plan to predict and achieve anticipated outcomes, monitoring the plan's execution helps prepare decision-makers for key decision points, often represented by branches and sequels in the plan. Branches and sequels are mini-contingency plans within the plan aimed at gaining on maintaining the initiative.

For attack plans, the primary emphasis is on maintaining and exploiting the initiative by predicting probable enemy responses and having predefined counters already prepared. For influence plans, predicting enemy messages and group insertions into the idea battlespace helps improve the speed and effectiveness of responses and can be used to be first in the arena with the message that defines subsequent discussion. For defense plans, execution monitoring helps determine when contingency plans should be launched, and whether those contingency plans were successful.

The decision to replan or not replan is a critical decision to make. This decision is best supported by having clearly measurable elements in the plan and especially the key decision points (including branches and sequels). If the conditions for branches, sequels, and predicted key events can no longer be achieved, it is time to replan.

Lastly, a process is required and a responsible agency needs to be assigned to monitor the success or failure of the planning factors and frequently update the values, categorizations, and qualifiers of the planning factors.

References

[1] Murphy, Colonel, "Quotes attributed to Colonel Murphy of Murphy's Law," http://www.military-quotes.com/ April 2006.

[2] Liddel-Hart, Sir Basil H., "Sir Basil H. Liddel-Hart quotes," http://www.military-quotes.com/, April 2005.

[3] Murphy, Dennis M., "War College and DoD Partner to Study Network Centric Warfare in OIF," *U.S. Army War College Alumni Newsletter,* Fall 2005.

[4] Bourchet, "General Quotes," http://www.military-quotes.com/ April 2006.

[5] Department of Defense *Dictionary of Military Terms,* August 31, 2005, http://www.dtic.mil/doctrine/jel/doddict/data/s/04806.html.

[6] Schwarzkopf, General H. Norman, *It Doesn't Take a Hero,* New York: Bantam Books, 1998.

[7] Parks, W. Hays, "The 1954 Hague Convention for the Protection of Cultural Property in the Event of Armed Conflict," http://www.kakrigi.net/manu/ceip4.htm.

[8] Unnamed Senior Defense official, "Briefing on Human Shields in Iraq," Global Security.org, February 26, 2003, http://www.globalsecurity.org/wmd/library/news/iraq/2003/iraq- 030226-dod01.htm.

[9] "Transcript: Confession by accused Jordan bomber," CNN online news, November 13, 2005, http://www.cnn.com/2005/WORLD/meast/11/13/jordan.confession/index.html.

[10] Halaby, Jamal, "Iraqi Woman Admits on TV to Jordan Attack," Associated Press article carried on Brietbart site, November 13, 2005, http://www.brietbart.com/news/2005/11/13/D8DS0JQO6.html.

[11] Associated Press, "Angry Jordanians to al-Zarqawi: 'Burn in Hell,'" November 10, 2005, http://www.msnbc.msn.com/id/9979747/.

[12] Nasr, Octavia, "Unverified Zarqawi Audio: Jordanian Civilians Weren't Targets," CNN.com, November 18, 2005, http://www.cnn.com/2005/WORLD/meast/11/18/zarqawi.jordan/index.html.

[13] "Bomber Confession Shocks Jordan," CNN.com, November 14, 2005, http://www.cnn.com/2005/WORLD/meast/11/14/jordan.blasts/.

[14] Clausewitz, Carl von, "Carl von Clausewitz quotes," http://www.military-quotes.com/, March 2006.

9
Planning Tools and Technologies

> *The users want a tool that requires only one card input, runs in a nanosecond, and generates a one-page output with all of the political, military, economic, and social implications worldwide.*
> —John Shepherd, 1981

Chapters 1 through 8 have presented concepts and mental models useful for information operations planning. This chapter describes a range of planning processes and tools, primarily emphasizing the planning tools that have implemented a large number of the planning concepts described in this book. Some of these tools were developed independently, but many have since become incorporated into the basis of DoD's next IO planning tool, the Information Warfare Planning Capability (IWPC) in preparation for its follow-on, the Information Operations Planning Capability-Joint (IOPC-J).[1] Along with the section topics below, this chapter briefly explains which of the previously described concepts have been incorporated into IWPC and other planning tools.

---

1. In 2003, the Department of Defense recommended the Information Warfare Planning Capability (IWPC) to become the basis for the Department of Defense's next IO planning system. The DoD IO Roadmap Recommendation 26 stated [1]: "The Air Force currently sponsors an IO planning capability. DoD should expand the Air Force's Information Warfare Planning Capability (IWPC) into a standardized IO planning capability at the joint level. This capability will serve as a suite of automated data analysis and decision support software tools designed to facilitate IO planning by Combatant Commanders. It will enable users to accomplish intelligence preparation of the battle space, Develop IO strategy and candidate IO campaign targets, plan IO missions, and monitor and assess execution."

Section 9.1 describes the planning process and efforts underway to provide semiautomated support to planning. This section also presents why a semiautomated planning process is more desirable than a fully automated process.

Section 9.2 describes a number of recent planning tools developed and fielded for IO and effects-based planning. Many of the separately developed IO planning tools have now been incorporated into IWPC as a precursor to developing the Joint version, IOPC-J.

Section 9.3 presents planning technologies explicitly focused on course of action (COA) planning, evaluation, comparison, and selection. As presented in Chapter 5, IWPC appears to be the first planning tool to explicitly distinguish between top-level COAs and employment option COAs. This distinction has helped alleviate previous conflicts and helped clarify the tools best applied by each planning cell.

Section 9.4 presents tools for predicting outcomes or effects in support of planning. This section discusses the pros, cons, and applicability of modeling and simulation at various levels of resolution to different planning echelons and types of planning.

Section 9.5 discusses the need for advanced adversarial reasoning technologies, as well as technologies that represent the reasoning of neutrals and other groups whose allegiances may shift over time.

9.1 Semiautomating the Planning Process

9.1.1 Why Semiautomated Planning Is Needed

As the military and interagency operating environments have become more complex, so too has the world of the military planner. For the U.S. military in particular, there is a wide range of planning considerations beyond how to destroy things. The emphasis in today's world is upon how to achieve the mission while destroying as little as possible and minimizing collateral damage, especially civilian casualties. Moreover, many of the current military planning considerations involve primarily nonmilitary objectives, such as peacekeeping, stability operations, and nation-building.

As we presented in Chapter 3, the complexity of information operations tends to be greater than for traditional military operations. A much wider range of attack, defense, and influence options are available, many more desired effects are possible, and there are many more tradeoffs to consider and a much larger decision space to deal with in terms of space, time, and enemy reactions. Due to the increased complexity facing military planners, the following two trends are logical solutions to this problem: distributed planning and semiautomated planning.

Distributed planning was introduced in Chapter 5 and will be described in more detail in Section 9.2. Due to increased complexity, it is difficult to staff every combatant command planning cell with an expert in every field. Therefore, reachback planning support is being applied at an ever-increasing rate to ensure that complex factors and tradeoffs are being adequately addressed during the planning process.

Semiautomated planning occurs when computer hardware and software are used to assist the planners in their planning. For example, if sufficient knowledge has been captured in a planning program, the program could recommend options to the planner. The Course of Action Support Tool (COAST) in IWPC presents a ranked list of options to the planner when the planner is seeking *employment options* to achieve a desired effect against a targeted Red activity. In a similar manner, Section 9.5 discusses adversarial reasoning technologies that are intended to recommend options to disrupt enemy activities.

9.1.2 Semiautomated Planning Versus Automated Planning

Note that semiautomated planning is not the same as automated planning. Automated planning is where the computer comes up with the whole plan or the overall plan and presents that plan to the planner for review. The author is not a supporter of fully automated planning for two reasons. First, a fully automated planning tool that works in real-world settings has not yet appeared on the scene. But the second and more important reason is that *automated planning makes the side that uses it very predictable.*

While it is true that one can add random factors to the recommended plan to try to make it less predictable, the context and categorizations that the software uses to develop the automated plan are fundamentally too rigid for real-world military planning. In order to be mathematically tractable and computationally manageable, automated planning tools need to make a wide range of simplifying assumptions that literally paint them into a box that is a subset of the space of all possible options. As a result, the automated planning side becomes predictable, and therefore an opponent can easily set up all sorts of unexpected situations that the automated planning system is not designed to handle.

Warfare and other military operations are still an art as well as a science, and artificial intelligence routines have not developed to the point of being able to out-think a human in all but the most rigidly constrained environments, such as chess or checkers. Even in computer gaming, which usually entails a wider range of options than chess or checkers, AI programs are usually a poor match against a human player. This human advantage over AI is even greater in the real world, where even more options are available to contestants than within a PC game. Therefore, the focus of this chapter is on those semiautomated planning tools that exist and are evolving to help address the IO and other complex military planning environments.

9.1.3 Human and Machine Strengths and Weaknesses

Semiautomated planning processes apply the strengths of humans and the strengths of machines, and try to avoid the weaknesses of each. For example, machines are very good at bookkeeping functions, such as storing, retrieving, and cross-referencing large amounts of data. This correct bookkeeping advantage also applies well to version management, configuration control of a plan, who is authorized to access which parts of a plan, who is currently editing which parts of a plan, and other collaborative and distributed planning features. This is one reason why machine-to-machine data exchanges are becoming more popular to help avoid human transcription errors in copying data from one machine or format to another.

The second area where machines excel is in performing a large number of repetitive tasks, such as sorting data; running algorithms; running computer models; formatting and displaying information in tables, graphs, or 2D or 3D displays; and otherwise doing the grunt-work tasks. Machines are very good at doing what you tell them to do, assuming that what you tell them to do is what they are designed to do.

Humans, on the other hand, still have a significant advantage over machines in terms of the ability to prioritize and ignore data, quickly identify key or critical factors, imagine what the enemy is thinking, and develop coherent and feasible courses of action that address a number of explicit and implicit tradeoffs simultaneously. As a result, analyzing the mission, developing a strategy, deciding upon the commander's intent, developing courses of action, selecting the military deception plan, and other cognitive-intensive tasks should be retained by humans and not delegated to machines. The machines should be used to store and disseminate the results of these human-performed tasks. For example, many of these human tasks may be assisted by machines, primarily in terms of visualization, but also in terms of presenting a prioritized list of available options when the situation (as perceived by the human) is fairly well defined so that machine can look up that situation in its memory.

9.1.4 The Need for Visualization

Visualization is usually the first choice for the planner when the situation is not yet well defined [2]. When the commander is attempting to comprehend the situation (including situation awareness, the space of available options, and the tradeoffs and probable outcomes for each of the options), the ability to display these related factors in a fairly understandable format is critical to the decision-maker. For example, the DARPA/JFCOM Integrated Battle Command Project recognized that the ability to visualize the complex space of diplomatic, information, military, and economic (DIME) actions applied to political, military, economic, social, information, and infrastructure (PMESII) features of a nation would require an advanced visualization capability [3].

Visualization is particularly important when the situation is unknown. The human mind learns by comparing and contrasting something new to something already known—determining what is similar versus what is different about this new situation. When the situation is unknown, the human mind benefits substantially from displays that show what is known and unknown, which enables the mind to identify patterns not readily apparent to machines.

The combination of the eye and mind is a great two-dimensional analyzer that can quickly grasp relationships across two dimensions. Machines tend to think in a linear sequence, and therefore require more time to determine the same thing the human does in an instant. As machines have become faster, the lag time between the human and the machines has reduced significantly, but when faced with unknown or unexpected situations, the human mind-eye combination still adapts much more quickly than does a machine.

9.1.5 The Need for Evidential Reasoning Tools

As an alternative to simply displaying the data and letting a person attempt to infer what is happening from the raw data, there are now evidential reasoning algorithms that help interpret the raw data by comparing it to contextual information. For example, if the context is known to be radio transmission during naval combat operations, and the word "paint" appears, it is more likely to be referring to using radar to "paint a target" than to "giving the ship a new coat of paint." A set of context-defining rules and terms are used to quickly sift through huge quantities of raw data to help perform basic reasoning that would take humans hours to perform. The results can be a fairly good set of reasoned inferences, assuming the initial context definitions were a close match to the real-world situation.

These inference algorithms build off contextual knowledge captured from a large number of subject matter experts to help the commander identify patterns in new situations. While still in their infancy, such algorithms show some promise. The way these algorithms work is to apply contextual data that filters less relevant data (based on the rules of the algorithm) and highlights anticipated relationships between data. The end result is that the machine can mine many more implications from raw data more quickly than a human because the meaning (or meaninglessness) of the data has already been defined by the humans who have placed that information into the algorithm and its rule set.

The good news is that these advanced algorithms could provide useful information not obvious to a human facing situations with many partially observed factors. The bad news is that the way these algorithms work—by building on assumed definitions for relationships of interest between data items—can lead the user out on an evidential limb where everything is concluded logically but may be completely wrong. Once again, the human mind reviewing the trail of evidence should have some understanding of how the algorithm builds its basis of evidence so the human can identify what the machine is likely to miss and which corner it is likely to paint itself into.

9.1.6 Planning Process Workflows

The U.S military commonly uses two planning processes: the Joint Operations Planning and Execution System (JOPES), and the Military Decision Making Process (MDMP). JOPES was intended for use in planning Joint operations, and therefore focuses on the theater level of planning. Figure 9.1 shows the different steps involved in deliberate and crisis action planning, as described earlier in Chapter 3.

Furthermore, the *Joint Staff Officer's Guide 2000* lists the theater estimate process in JOPES as [4]:

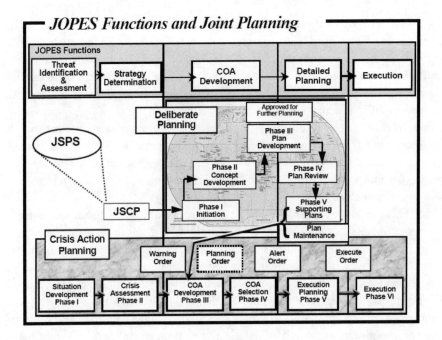

Figure 9.1 The JOPES functions and Joint planning.

1. Mission
   a. Mission analysis
   b. Mission statement
2. Situation and Course of Action
   a. Situation analysis
   b. Courses of action analysis (develop each COA, where each must be adequate, feasible, and acceptable.)
3. Analysis of opposing COA (likely effects of opposing COAs on friendly COAs)
4. Comparison of own COA (comparing friendly COAs)
5. Decision (pick one COA to implement as the plan)

Joint planners tend to use the JOPES process because it is commonly understood and appropriate to the theater and operational levels of planning. Planning tools that follow this planning process will be familiar to planners that use this process. For example, the user interfaces in IWPC were designed to follow the JOPES planning process, as it was intended to plan operations at the theater and operational levels.

At the same time, the MDMP is another similar planning process that originated with the Army, but has also become more popular in the Joint community. The U.S. Joint Forces Command recently published *The Commander's Handbook on an Effects-Based Approach to Joint Operations*, which uses the MDMP as the starting point for effects-based planning. HQ 1st IO Command Land also bases their planning on the MDMP.

The sequence of planning steps in the MDMP is [5]:

- Receipt of mission;
- Mission analysis;
- COA development;
- COA analysis (war-gaming);
- COA comparison;
- COA approval;
- Orders production.

MDMP also tends to be applied by subordinate commands down to the tactical level, especially by land forces.

Table 9.1 shows that there is not much difference between the two planning processes, which are simply different ways to slice or categorize all of the

Table 9.1
The JOPES Planning Process Compared to the MDMP Planning Process

JOPES Planning Steps	MDMP Planning Steps
(1) Mission	Receipt of mission
a. Mission analysis	Mission analysis
b. Mission statement	
(2) Situation and Course of Action	
a. Situation analysis	
b. Courses of action analysis (develop each COA, where each must be adequate, feasible, and acceptable)	COA development COA analysis (war-gaming)
(3) Analysis of opposing COA (likely effects of opposing COAs on friendly COAs)	COA comparison
(4) Comparison of own COA (comparing friendly COAs)	COA comparison
(5) Decision (pick one COA to implement as the plan)	COA approval
	Orders production

steps necessary in planning. Similar tables could be developed for the JOPES deliberate planning process and the crisis action process.

Both the JOPES and MDMP planning processes are used by different military organizations. However, the planning tools developed for one will sometimes be unfamiliar to those using the other type of planning process because the slicing or categorization of the steps is slightly different. For example, when HQ 1st IO Command Land reviewed IWPC as a tool to support its planning, it rightly required that a new set of GUIs be developed so that the planners could use and follow the MDMP. In a similar manner, Special Operations Command was used to the SOFTools user interfaces to support planning at the mission level, and would like to continue to use that interface for their planners.

In general, planning tools need to have graphical user interfaces that walk the user through the steps with which they are familiar, rather than force them to switch to the other planning process. The development of new GUIs and adding new data structures to an existing planning tool like IWPC is much less expensive than developing a new planning tool from scratch. Note that the semiautomated planning tools should be capable of supporting either planning process, and to be able to translate planning artifacts from one process to the other.

The benefit of common planning processes is that everyone is familiar with the steps and can perform their planning steps efficiently. However, different planning cells tend to prefer to use slightly different (but not radically different) planning processes that they have found better suited to planning their operations. For ease of use, it is usually better for the planning tools to adapt to these planning processes than vice versa. That means that there either needs to be a separate planning tool for each one, or a common planning tool to translate the data between the different processes so that each planning cell is on the same sheet of music. Each planning cell can follow the process it prefers, but the planning tool(s) need to be able to seamlessly translate their planning artifacts for each cell. This is discussed further in Section 9.2.2.

9.2 Semiautomated Planning Tools

Just as no man is an island, no planning tool should be an island when planning in today's Information Age. The planning tools need to be integrated into an overall system of people, equipment, processes, workflows, information systems, communications systems, and organizations. Planning cells require:

- A planning process and workflow commonly understood by all participants (as described in Section 9.1.6);
- Personnel trained in the process and tools (see Chapter 10);
- The planning tools themselves (see this section);
- Execution monitoring and replanning tools integrated with the planning tools (see Chapter 8);
- Real-time feeds from real-world sensors and information systems (see Chapter 8);
- The data and databases to support the planning tools (see Chapter 10);
- Communications and collaboration between various planning cells and their tools (see this section);
- Interfaces to additional analysis tools at the reachback centers (see Chapter 8);
- A feedback loop to review and evolve the accuracy of the planning factors over time (see Chapter 8).

When a planning tool is fully integrated into a system with all of these factors, it has the best chance of contributing to effective and efficient planning. Planning tools that integrate with and "play well" with existing planning tools and the JOPES and MDMP planning methodologies will be accepted more readily than those that do not.

9.2.1 Current Planning Tools

A few years ago, the planning tool used most often by the U.S. military was Microsoft Office—Word, PowerPoint, and Excel [6]. Even today, Microsoft Office is still extensively used in most planning cells, especially for preparing status briefings or decision briefings. The advantage of using this tool suite is that every planning organization has it, and little or no training is required for the staff to use this suite. The bad news is that absolutely everything has to be done manually because these tools are intentionally so simple.

Because planning with these tools is so manpower intensive, planners often choose to take shortcuts, such as only developing a single course of action rather than the recommended three for purposes of comparison. If only one COA is developed, there is nothing to serve as the basis for the military deception plan, which is one reason why deception plans are not always created.

More can and should be done to get past the highly manpower-intensive planning of solely using Microsoft Office products. For example, during Joint Expeditionary Force Experiment '04, the planners using dedicated planning tools were able to develop IO support to the overall strategy, and to develop the target list, much more quickly than the cells still using only spreadsheets [personal communication with Tim Autry, May 12, 2006].

This is not to imply that there is a single planning tool that is applicable to every planning cell. There is no one-size-fits-all tool due to the differences in the echelon of the planning cell, its planning horizon, the types of assumptions required, and the decisions being made at that echelon.

9.2.2 The Echelons of Focus of Each Planning Tool

Every planning cell resides at a particular echelon and has a particular role to play in the planning process. Some planning cells plan the next operation, while others focus on the operation after next. Some planning cells focus on defining and monitoring the achievement of measurable plan objectives, while the warfighters monitor the execution of the plan for the approach of decision points, such as branches and sequels.

For example, the Special Operation Forces planning tool suite (SOFTools) is specifically designed to support the planning and execution monitoring of SOF missions. In addition to a dynamic synchronization matrix, the tool can be used during execution monitoring to track the call signs used to check off and track whether the next step in the plan has been accomplished. This feature is very useful to the SOF mission planner and warfighter during execution, and the approach may be applicable to other types of tactical mission planning and execution, such as intelligence planning and collection. SOFTools was not originally designed for theater-level planning.[2]

So the first question to ask of any existing or developing planning tool is: What are the echelon and cell(s) this planning tool is designed to support? If the answer is "every planning cell," then the tool proponent does not understand the planning environment and its requirements.

The tool development process requires the design of the overall tool (or each specific component tool of a suite) to focus on the features of interest to a particular planning cell—the decision it is supposed to make, the products it is supposed to develop, the assumptions it needs to make, and the context in which it is making these preparations. All of these factors influence design decisions that tailor the tool for one planning cell, but also limit its applicability to another planning cell.

For example, the Air Operations Center (AOC) has five planning divisions, and each division has from one to three planning cells, each tasked with different responsibilities. Due to these different responsibilities, each cell uses a subset of the planning tools available to perform their planning tasks, as shown in Table 9.2.

Joint headquarters and other component Service headquarters have their own set of planning cells, and sometimes use similar tools, while others use different tools. In addition, some planning tools require the user to have a substantial body of knowledge in order to use the tool effectively. As a result, some planning tools are more appropriate for use by reachback centers, where such specialized skill sets are more readily available.

This need to tailor each planning tool to a particular planning cell or set of cells is one reason why suites of planning tools are becoming popular. Rather than develop a single planning tool per planning cell, some common features and functionality can be leveraged across a range of tools. IWPC is one such suite of tools, where each tool is intended for use by one (and sometimes more than one) planning cell. Even with the underlying common features, however, each individual planning tool is tailored to the needs of a specific planning cell. Although the data is stored in a common database, the data display and data entry options presented to a given planning cell are specifically tailored to the needs of each planning cell. Being able to exchange information among the various planning tools is essential in the modern planning environment. Whereas suites of tools can share data directly between their components, tools outside the suite usually use XML or some other data exchange protocol.

---

2. Note that SOFTools has recently been modified and adapted to the Integrated Battle Command Project to support interagency planning cells, but does not use the tactical level call-sign tracking feature for its new theater-strategic planning application.

Table 9.2
AOC Planning Divisions, Planning Cells, and Selected Planning Tools per Cell

Planning Division	Planning Cell	Selected Tools
Strategy division	Strategic plans	Course of Action Support Tool (COAST)
	Strategic guidance	Collaborative Planning Tool (CPT)
	Ops assessment	Enhanced Combat Assessment Tool (eCAT)
Combat plans division	Target effects team	Joint Targeting Toolbox (JTT)
	Master air attack planning (MAAP)	Theater-Battle Operations Net-Centric Environment (TBONE)
	ATO production	TBONE
Combat ops division	Current day's ATO (several cells)	TBONE, Web-Enabled Execution Management Capability (WEEMC), Data Link Automated Reporting System (DLARS), Advanced, Field Artillery Tactical Data System (AFATDS)
Air mobility division		TBONE et al.
ISR division	Analysis correlation and fusion (ACF)	CPT and eCAT

Even with this increased degree of flexibility, however, IWPC is still not the only planning tool required, even for planning just information operations. Since IWPC's focus is on the operational and theater levels of planning, more detailed planning will be required by cells below it, and some higher-level interagency planning tools (such as the Integrated Battle Command) will be required for higher-echelon interagency planning. For example, the PSYOP planning module currently in IWPC is designed only to capture the top-level strategic/ theater-level elements of a psychological operation and how it fits into the overall plan. Much more detail needs to be added to planning this operation before it would be considered adequately planned, including additional details of desired effects, the type and timing of the delivery and effects, the actual delivery mechanism and limitations, and the nuances and biases of the intended audience.

With this inherent limitation in mind, the IWPC program has developed the request for analysis (RFA) and request for detailed planning (RFDP), concepts introduced in Chapter 5.[3] Designing IWPC with the RFA feature ensures two desirable results. First, it provides the mechanism to ensure a complete audit trail for the complete planning process, including the participation of the reachback centers and their resulting advice. (The planning cell requests that the

---

3. Although originally designed and developed for the Course of Action Support Tool, the RFA feature has recently become a service available to all tools in the IWPC tool suite.

reachback center analyze a particular question, and the answer is reviewed for inclusion in the plan, while IWPC maintains a complete audit trail of the transaction.)

Second, having the RFA capability encourages the separation of supporting planning tools that provide additional detail from the main theater-strategic or operational-level planning tool. This separation helps ensure that developers do not overload a single tool with too many built-in capabilities. This overload problem has occurred in many planning tool developments, causing the main tool to become too cumbersome to use. The availability of the RFA capability helps keep the tools that need to be with the main planning tool appropriate to the primary users of that suite, and defines a clear system boundary that can be used to determine which new tools need to become part of the main tool or serve as a support tool. The additional detailed, tactical, or employment option planning tools need to remain as separate entities, but capable of exchanging data directly with the operational-level planning tool without having to reenter the data manually.

9.2.3 Current IO Planning Tools

Like most other planning tools, IO planning tools often started out as PowerPoint briefings, Word documents, and Excel spreadsheets. As IO planning procedures evolved, some of these spreadsheet tools became more elaborate, such as the Information Operations Navigator (ION) planning tool used by the Joint Information Operations Warfare Center (JIOWC). This tool has been used by IO planning support teams deployed to the Combatant Commander staffs during conflict. However, the simplicity of a spreadsheet planning tool was found to be insufficient to support IO planning, especially as the requirements for collaborative planning continued to grow. Since the ION tool already had a following for its use, it was integrated into the IWPC planning suite as another component tool. Now ION can share and exchange data with the other members of the IWPC tool suite, thereby gaining substantially more capabilities. The JIOWC has agreed to use IWPC as its IO tool of choice in compliance with the DoD IO Roadmap.

Over the last few years, a number of other IO planning tools have been made part of the IWPC tool suite. The set of IO planning tools that are a part of IWPC at the time of this writing is presented here:

- Collaborative Workflow Tool (CWT) for tracking planning workflow progress;
- Analyst Collaborative Environment (ACE) to support intelligence analysis and situation awareness;

- Course of Action Support Tool (COAST) for COA development, evaluation, comparison, and selection (described further in Section 9.3.1);
- Collaborative Planning Tool (CPT) for elaborating the details of the plan once a COA has been selected;
- Information Operations Navigator (ION) for planners familiar with the legacy ION planning tool;
- Interactive Scenario Builder (Builder) for electronic warfare planning;
- TEL-SCOPE for communications link target selection (communications network target data, analyzed and selected in TEL-SCOPE, appear as targets in the IWPC database);
- Target Prioritization Tool (TPT) for modeling space and terrestrial communications;
- Enhanced Visualization (eViz) for geographical displays of the plan elements;
- Enhanced Synchronization Matrix (eSyncMatrix) for scheduling and plan deconfliction;
- Enhanced Combat Assessment Tool (eCAT) for planning how actions and their desired effects will be measured for progress and success;
- Execution Monitoring Tool (EMT) for tracking the progress of the plan compared to real-world results and the status of decision-point criteria;
- XML Briefing Composer (XBC) to help build plan review briefings automatically.

One of the primary advantages of the IWPC tool suite is its ability to successfully support collaborative, distributed planning. IWPC uses an Oracle database as a central repository, applying Oracle's row-level security to the different elements of the plan. This allows each planner to work on developing his part of the plan in parallel with all of the other planners. The planner that is currently editing a single plan element (objective, task, or effect) is the only person that can write to that part of the plan (represented in the database), while all other interested parties can see what was entered most recently. This approach helps avoid the risk of two different planners overwriting each other's work. As a result, IWPC was declared to be the only truly collaborative planning tool that was used during JEFX '04 (author conversation with Tim Autry, June 2006).

The collaborative planning capabilities of IWPC are essential to support distributed planning. The planners connected to the central database or its supporting data provision services use the inherent collaborative features to do their work. For those more distant and not connected to the IWPC servers and services, the request for detailed planning (RFDP) feature is used instead. This feature ensures that the required plan elements and rationale from the supporting planner is stored as part of the complete audit trail for the plan's development.

Another feature essential to modern planning is the ability to provide a complete audit trail of the planning considerations—what was selected and why, what was not selected and why not. As described in Chapter 11, an accusation is a powerful weapon in the idea battlespace. In the past, the United States has had to recreate what led to certain decisions and plans in order to defend itself against accusations in arenas such as the United Nations International Court of Justice. Having a complete audit trail already available as an automatic product of the plan development process is an essential survival tool in the modern world.

As noted earlier, care must be taken not to overload IWPC, or any planning tool, with too many component tools, especially those designed for use by reachback centers for detailed analysis, or by more tactical level planning cells developing employment planning and plan elaboration. This is particularly true of many of the models and simulations designed or proposed to provide detailed analysis to the IO planning process, as described further in Section 9.4. Understanding where IWPC should end and other tools begin is an important consideration for whether any tool should become an integral part of IWPC or whether it should simply exchange data with it. Maintaining that system boundary will help ensure a long life for IO planning tools suites and a consistent planning environment for the IO planner.

9.3 COA Planning Technologies

9.3.1 The COA Development Process

Course of action (COA) development, evaluation, comparison, and selection are planning steps in both the JOPES and MDMP processes. The JOPES process explicitly compares the effects of enemy COAs on friendly COAs, while those effects are accounted for during the MDMP COA analysis (war-gaming) step.

The JOPES process includes defining friendly and enemy courses of action (COAs), evaluating each one, comparing the various COAs to each other, and selecting one for more detailed planning and execution. Usually, two or three COAs for each side are compared, and keeping all the details straight can be a significant bookkeeping challenge. Moreover, the methodologies used to evaluate and compare COAs can have a significant impact on which COAs are chosen. There is also a tradeoff between the time available to plan and the amount of detail required to fully understand the advantages and disadvantages among the various COAs.

Each COA may contain its own contingency elements, such as branches and sequels (defined in Chapter 8). A branch defines an action to be taken if certain conditions are met, while a different action will be taken if different conditions occur. A sequel defines an action that awaits a certain set of conditions to be met before occurring. Note that defining and keeping track of the branches, sequels, and conditions for each COA can further add to the bookkeeping burden.

9.3.2 Types of COAs

Over the years, a number of tools have claimed to be COA planning tools, and some were more applicable than others to particular planning cells. To date, the author has not been able to identify any COA planning tools prior to COAST that distinguished between the different levels of COAs that need to be developed in a given plan. The top-level COAs compare Blue alternative COAs to Red alternative COAs to identify their probable outcomes and select the best COA to be the top-level plan. Once that top-level COA has been selected, additional employment (or how-to) alternative courses of action need to be developed, as there may be many ways to implement a given plan. Depending upon the number of echelons involved in an operation, there may be many subordinate COA comparison sets that need to be prepared, where each subordinate echelon determines the how to accomplish for the higher echelon's what to accomplish.

For example, in traditional Army planning, the Corps headquarters would develop COAs, and select one for elaboration by the Division headquarters, the next echelon down. The Division HQ would develop its own COAs, select one, and pass that down to its subordinate commands (the Brigade HQs) to elaborate upon. These Brigade HQs would repeat the process for each of its subordinate commands. Thus the subordinate level HQ developing COAs to determine the best how-to plan from the higher HQ's selected plan (based on its COA development) was an inherent part of traditional military planning processes. However, the previous tools developed to support COA planning did not do well at coordinating and passing data from one COA planning HQ to another.

Information operations planning is subject to the need for a similar hierarchy of increased planning detail for the how-to implementation for any given plan. Part of the problem was that the U.S. Air Force has focused on a single Air Tasking Order (ATO) to orchestrate the activities of every aircraft every day. While this is a good end product of the planning process, it usually did not account for the fact that different levels of planning had to be performed throughout the hierarchy for the final product to be adequately prepared.

For example, if the mission is to preclude an enemy invasion across the border, one of the component objectives could be to delay the movement of enemy forces. There are many ways to accomplish that, including directly attacking the forces, reducing their flow of supplies, or disrupting their command, control, and communications (C3) capabilities. These three alternatives represent three COAs that need to be developed, evaluated, compared, and selected. Let's assume that one of the political constraints is no kinetic options before hostilities begin. That reduces the options to reducing the flow of supplies and disrupting enemy C3.

Let's assume, therefore, that disrupting enemy C3 is the selected how-to COA. The next question is: What is the best way to disrupt enemy C3 without using kinetic options? Many options are available, and each can be compared until one or more is finally selected. This process of adding more detail continues down the hierarchy until one combination of options is selected for implementation. These selections then roll back up the hierarchy to the top-level plan. Thus in the case of the air combatant commander, the ATO for the day was based on performing COA comparisons all down the command hierarchy and back up again.

9.3.3 The Course of Action Support Tool in IWPC

This process of elaborating, evaluating, comparing, and selecting employment COAs to satisfy the top-level plan is built into the IWPC Course of Action Support Tool (COAST).[4] As mentioned earlier, it is unlikely that IWPC will actually be used in its current form at every tactical command level, such as an Army company, since it was not designed to support planning at that level. However, the ability to reach down many levels and include subordinate COA processes as part of the audit trail is a powerful capability for plan review, and for evaluating which planning factors need to be revised over time. For example, if the rationale for why certain options were selected and others were not is captured automatically during the planning process, it becomes far easier to identify problems with either the planning factors or the rationale in determining what went wrong and why.

As mentioned previously, the purpose of the automated audit trail is to store all such deliberations for future review and evaluation to help improve planning factors over time. None of the COAs is thrown away. Instead, every COA that is evaluated is stored, and each rejected COAs becomes a candidate for the basis of the deception plan for that echelon for that employment option.

Another first for COAST is its inherent ability to define and track the satisfaction or violation of political guidelines and constraints for any element of the plan. The higher command authority can define a standardized set of rules

---

4. General Dynamics is the prime contractor on IWPC and developed COAST.

of engagement (ROEs) that will be used to guide subsequent planning. This top-down guidance will help preclude the problem of making detailed plans that are quickly rejected by the approving authorities for being beyond what was expected. The planners and commanders at every appropriate level throughout the planning process can now determine at a glance which parts of the plan are likely to violate an ROE guideline or constraint, and thereby help focus attention on how to mitigate such effects or to select a different option. This capability provides the first quantitative and graphical representation of ROE restrictions on military operations, as described in Chapter 5. It also provides, for the first time, a graphical representation of the effects of the political guidelines and constraints on the options available to the military commanders.

COAST also developed an expanded set of target considerations that could be defined by category of target [such as the categories defined in the Military Intelligence Database (MIDB)], or for each specific target. In addition to the category of no strike, there are many other target considerations, such as the 13 categories shown in Chapter 5. Since no strike was a term created when kinetic options were the only ones available, the ability to define many different types of kinetic and nonkinetic types of strikes and their effects is essential to target planning in the Information Age.[5]

COAST, of course, was not the only COA tool developed over the last few years. For example, the Strategy Development Tool (SDT) was developed by ALPHATECH, Inc. (now part of BAE Systems). SDT has many similarities with COAST. Both SDT and COAST have a cause-and-effect type of display, as well as a similar database structure. The two tools were sufficiently compatible that desirable features of SDT were incorporated directly into IWPC by the developers of SDT as part of the AOC Strategy Development project [7].

The following concepts presented in this book have been incorporated into COAST to date:

- Blue COA development (top-level and employment option COAs);
- Red COA development (top-level COAs);
- Political (ROE) guidelines and constraints;
- Target considerations (being moved to CPT in the next development cycle);
- Cause-and-effect networks;

---

5. Note that when it was determined that the planning cells most likely to define target considerations would use the IWPC Collaborative Planning Tool (CPT), the decision was made to move the target considerations module from COAST to the CPT module in the next build cycle. Another advantage of having these planning tools part of the same suite of tools is that shifting capabilities from one product to another is more readily facilitated.

- Branches and sequels (both in planning and execution monitoring);
- Blue Capabilities Matrix;
- Kinetic and nonkinetic options integrated in the same plan;
- Red versus Blue top-level COA evaluation, comparison, and selection;
- Blue employment option COA assessment, comparison, and selection;
- Ranking COAs based on user-defined scores and ROE satisfaction;
- Red reactions and Blue counterreactions (see Chapter 8);
- Request for analysis (now expanded as a service to many IWPC tools);
- Request for detailed planning (for planners not connected to main database server or services);
- Execution monitoring;
- Idea battlespace Part I (groups, messages, message purposes, basic situation awareness, PSYOP task development wizard);
- Complete audit trail.

The following concepts presented in this book are still awaiting funding for IOPC-J to be implemented (although the high-level designs have been prepared):

- Military deception and OPSEC planning;
- Public affairs planning;
- Idea battlespace Part II (including a counterpropaganda model).

In addition, other tools have incorporated some additional concepts presented in this book:

- Integrated Battle Command has developed Phase I of the idea battlespace information and media model.
- For the Blue Capabilities Matrix, both competing teams on the Integrated Battle Command project have accepted the DIME-by-PMESII matrix as part of their applications.
- A variant of the cause-and-effect network was accepted by one of the teams competing in the Integrated Battle Command Project.
- Integrated Battle Command is including both kinetic and nonkinetic options in the same plans.

- A CND module called course of action process (COAP) has been developed for the Integrated Information Infrastructure Defense System (I3DS), but not yet implemented.

As these two lists demonstrate, some of the concepts presented in this book have been accepted and are already in use, while others are still being evaluated. As a result, a mixture of traditional and innovative ways to approach and perform IO planning is being applied in the real world.

To some degree, the COA and other semiautomated planning tool development described earlier has led to improvements in the doctrine. In addition to the ROE guidelines and constraints, the revised categorization of the desired effects has been fed back into IO doctrine houses for their consideration. For example, the Blue Capabilities Matrix and the cause-and-effect network were applied to the U.S. Joint Forces Command problem of how to visualize the operational net assessment database in support of EBP. The DIME-by-PMESII and modified CAEN representation tailored for their problem was a major success, which helped lead to IWPC being planned as part of the standard suite of tools in the Standing Joint Force Headquarters. This preliminary effort was then incorporated into the two competing teams of the Integrated Battle Command Project, since that project was intended to evolve EBP beyond relying primarily on the operational net assessment for its data.

In 2001, Wilkins and desJardins published an article with a list of the capabilities that were required to solve real-world problems: numeric reasoning, concurrent actions, context-dependent effects, interaction with users, execution monitoring, replanning, and scalability [8]. IWPC with COAST has all of these features to some degree, including numeric reasoning (subjective scoring of Blue and Red COAs, COA comparisons, the TEL-SCOPE model, and the TPT model all use numeric reasoning), concurrent actions (in CAEN and eSynch matrix), some context-dependent effects (in the Blue Capabilities Matrix), definite interaction with users, execution monitoring (the Execution Monitoring Tool), replanning (see Chapter 8), and scalability (applicable to multiple levels of planning, but not every level). Even so, the topic of numeric reasoning tools is so large that much more can be done in this area. As described previously, the intent is to prevent IWPC from becoming so large that it becomes too cumbersome to use. Therefore, IWPC is designed to interface with other more detailed numeric reasoning tools, as described later in this chapter.

9.3.4 Other COA Development Tools for EBP

Effects-based planning (EBP) supports COA development and selection for effects-based operations (EBO). The Air Force defines EBO as [9]: "a methodology for planning, executing and assessing operations to attain the effects required to achieve desired national security objectives." The Military Operations Research Society Symposium on Analyzing Effects Based Operations [10] described two main features of EBO:

1. Effects-based operations challenge us to move from an era of increasing jointness to an era of meta-jointness that integrates the DoD's actions into coherent sets of actions that involve a broader set of participants [e.g., interagency and coalition partners, international organizations (IOs), and non-governmental organizations (NGOs)].

2. Effects-based operations require both greater knowledge and greater capability to deal with uncertainty than traditional military operations.

According to [10], "EBO coordinates sets of actions directed at shaping the behavior of friends, foes, and neutrals, in peace, crisis, and war." In other words, EBO planning reaches beyond just military planning into the realm of interagency and international organization planning to coordinate or at least deconflict the efforts of all friendly actors in a region. The workshop also concluded with the need for the community to [10]:

Develop a "tool chest" to support EBO analyses that includes easily manipulated, specialized modeling and simulation tools, computational social science tools, data mining, colored Petri nets, neural networks, and specialized tools developed in particular application arenas (e.g., counter-terrorism, persuasive communication, economics). This tool chest should be assembled in evolutionary fashion, creating a core capability from "best of breed" products and refining and expanding the tool chest to reflect user feedback and the results of research.

To meet this requirement, many of the new COA planning tools have been formulated to support effects-based planning (EBP). We have selected five examples to discuss here. Note that the first two were designed primarily to address EBP against an adversary in order to modify their behavior, while the latter three are examples in which the effects could be intended to be beneficial to the recipient, such as in nation-building operations. The fact that EBO can result in beneficial effects to friendly participants is sometimes missed in the discussions of EBO, which tend to focus solely on the disruption of enemy activities and modifying enemy behaviors.

Our first example is the Strategy Development Tool (SDT), which was developed by ALPHATECH, Inc. (now part of BAE Systems). The primary intent of SDT is to provide commanders with "with some understanding of how their potential courses of action might play out" [11]. SDT's COA development process helps the planner define and display of the cause-and-effect relationships among actions and effects, and uses a *Bayesian Belief Net* approach to assess or war-game the anticipated outcomes of each COA.[6] Figure 9.2 shows a sample screen from the SDT.

A second example is a systems dynamics (SD) model, designed to be a tool for EBO analysis and described by John Byrne. According to Byrne, SD models are best used in EBO analysis when limited to answering specific questions, such as evaluating a hypothetical solution. In particular, SD models can be used to find long-term patterns and high-leverage decision points. At the same time, SD models can be linked to each other to represent the various phases of the campaign, including mobilization, deployment, bed-down, and combat missions. The graphical construction of SD code allows for the drill-down from the overall mission to the specific systems and their underlying technologies. Byrne concludes that [13]:

> Effects Based Operations (EBO) analysis requires simulation to capture the cause-effect relationships in logical interdependencies. This is particularly true when the problem is less linear and more dynamic, when the problem contains feedbacks and/or delays, when the question mixes dissimilar systems, when particular subsystems or technologies are of central interest, or when the problem contains human factors. Solutions to these types of problems need the support of a systemic architecture discipline and are benefited from an intuitive quick means of developing computer simulations.

A third example is a nation-building model from the Air Force Institute of Technology (AFIT), which is also a systems dynamics model [14]. After the primary combat missions are over, stability and reconstruction operations begin. This SD tool was designed to support decision-making analysis at a subnational, or regional, level. The SD model provided drill-down capabilities similar to those described by Byrne to investigate areas of concern for stability planning. This model applied two different measures: a probability of stabilization success, and a probability of stabilization failure. This model was tested with sample data from Operation Iraqi Freedom.

The fourth example that we present is an agent-based model. The Synthetic Environment for Analysis and Simulation (SEAS), developed at Purdue University, is intended to simulate "the non-military, non-kinetic, and non-lethal aspects of modeling and simulation—diplomatic, religious and social

---

6. SDT also has an adversarial reasoning model, which will be described further in Section 9.5. SDT's Bayesian war-gaming tool is sometimes described as a separate tool, called the Operational Assessment Tool (OAT).

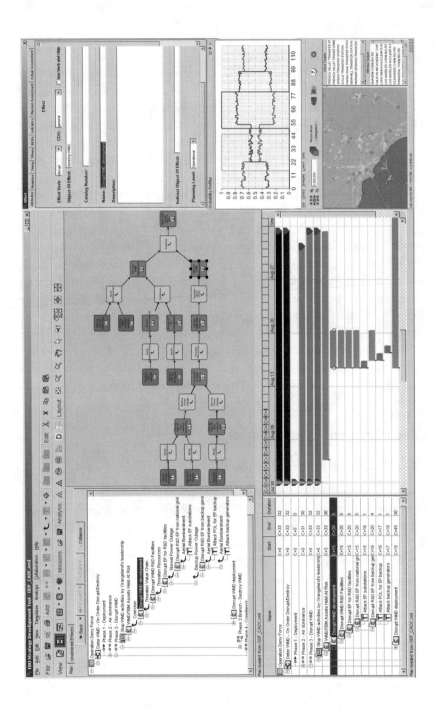

Figure 9.2 Sample screenshot of the Strategy Development Tool [12].

dimensions" [15].[7] U.S. Joint Forces Command has been applying SEAS to support their transformation experiments, using the model to represent diplomatic, economic, political, infrastructure, and social issues [16]. The model's agents represent either individuals, such as key leaders, or groups, such as terrorist organizations, the media, governments of a city or country, nongovernmental agencies (NGOs) and international corporations. Human participants in the experiment interact with each other and with the simulated agents in SEAS.

The fifth example is really two suites of models that include agent-based models, system dynamics models, and other model paradigms, each combined on a single planning platform. DARPA's Integrated Battle Command (IBC) Project has led to the development of two prototype suites of tools called LIVE and COMPREHEND, which will be used to simulate the interactions of all of the PMESII elements of a nation with each other, and the effects of DIME actions taken to change current states. For example, in LIVE, all the various types of models affect a partitioned set of PMESII variables, which are in turn used as inputs to appropriate models in the suite. The suite includes models of political, economic, and social interactions among the various actors, and their effects on each other and the PMESII state vector itself. As a result, the effects of planned actions can be estimated in the context of the complex space of interactions in the area of interest. While still in Phase I of development at the time of this writing, the two IBC suites of tools designed to assess the DIME and PMESII factors in a region of interest using a wide range of modeling paradigms appears to be a promising advance in the suite of tools available to support EBP [3].

9.4 Effects Prediction Technologies

9.4.1 Benefits of Prediction Technologies

When prediction technologies can adequately predict the likely range of outcomes of various conditions and actions, planners and commanders gain confidence in the planning tools and their products, which increases the likelihood that the mission objectives will be achieved. The JOPES planning process includes war-gaming each COA to see how it plays out. This is usually performed as a mental exercise and walk-through, although sometimes military simulations have been used to test out plans when sufficient time to prepare the simulations was available.

For example, prior to the ground assault during Operation Desert Storm, the critical "berm breaching" operations of the Coalition forces were simulated in a wargame. What they found was that the berm breaching was taking longer

---

7. Note that there are a number of models named SEAS, such as the agent-based System Effectiveness Analysis and Simulation developed for the Space and Missile Systems Center.

than expected, as the units waiting to pass through started bunching up at the entrance and became a lucrative target for Iraqi artillery. As a result, the berm breaching plan was modified, the method by which units were called to the breach was better orchestrated, and the counterbattery plan was improved. The execution of the plan in the actual operation came off without a hitch [personal communications with Richard Rouse, November 1996].

In a similar manner, when planning for Operation Just Cause to liberate Panama, commanders rehearsing the mission found that some of their units were overextended and exposed to heavy enemy fire during part of their planned maneuver. By modifying the maneuver plan and having better on-call access to airborne fire support, the modified plan avoided the heavy casualties and delays when similar opposition was encountered during the actual operation [personal communications with Richard Rouse, November 1996].

Thus when the range of outcomes of plans and actions can be adequately predicted, planners have the opportunity to work out problems in the plan and correct them before actual execution. Like physical mission rehearsal, simulating a walk-through of the plan can help determine where problems will arise and whether the plan needs revision before actual execution.

In addition to the simulations or wargames described earlier, many other prediction technologies could benefit military planners. For an information operation, it would be useful to predict whether the desired effects are achieved at all or achieved on time, and whether there were any unexpected undesired effects. At the same time, however, most prediction technologies require a large amount of highly detailed information to adequately predict outcomes. And as described in Chapters 3 and 10, the amount of detailed information required to plan and simulate an information operation can be significant. Therefore, planners need to be aware of the tradeoffs associated with the time required to prepare, run, and analyze outputs of models compared to the resources available, and whether the outputs of the models and simulations provide sufficient valued-added to plan preparation and assessment.

9.4.2 The Tradeoffs of Modeling and Simulation Tools

Every prediction technology is designed to support some sort of decision, whether an investment decision, a military decision, and an IO planning decision, or some other decision. Good models and simulations could help decision-makers answer specific types of questions. There are no one-size-fits-all models or simulations, which is why intentionally different models are being applied in the military to support analysis, training, testing, experimentation, acquisition, and planning. For each type of decision, the level of detail provided by, and required by, the model can range from very little to quite a lot.

Just as planning tools have an echelon of focus, so do models and simulations. The following is a list of some of the model design factors that help differentiate models used for different applications:

- Who is the intended user set of the model or simulation?
- What is the decision process being supported by the model?
- What assumptions can be made when supporting this type of application?
- What data is required to feed the model?
- Is the input real-time data streaming or can it be prestored?
- What level of input data consistency checking is required?
- What is the time available to input the data?
- What is the time available to run the model?
- How many model runs are required per time period?
- How quickly can the output data be accessed by the user?
- How realistic are the outputs in terms of providing support to the decision at hand?

For example, models used in experimentation tend to examine future events 15 or more years hence. Thus, less detail is required because the world situation 15 years from now is highly uncertain, but on the other hand, such models need to be broad in scope to be able to address a wide range of global changes over time. Analytic models may require the ability to run a large number of times to answer a single question.[8] Moreover, analytic models are often required to be repeatable and adequately cover the space of options. In contrast, training models need to provide plausible results in one iteration, but these results must be plausible the first time because there is no chance to repeat the whole event. Planning models need to support command decisions because people's lives are often at stake.

The more complicated the subject being predicted, the more likely that a more detailed modeling or simulation representation will be required. However, every added variable in the model, every new tradeoff represented, every new input data requirement, and every new output display add to the overall complexity of the model or simulation and therefore increase the cost to develop and operate that model or simulation.

---

8. For example, the author coauthored a paper where we ran 800,000 cases of a simple stochastic model to demonstrate that anomalous results occurred frequently when others were claiming they could not be occurring at all [17].

Thus a number of competing factors in model design help determine the type of prediction model required, what it will cost to develop the model, what it will cost to provide care and feeding of the model (e.g., feed it data, set up scenarios, and run parameters), what it will cost to process and comprehend the output, how long it takes to prepare and run the model, and how good a prediction (or range of outcomes) it provides to the decision-maker.

In general, the more detailed the model, the more expensive it will be in terms of development and care and feeding (operations). Also in general, the combination of topic breadth (how many factors it considers) and depth (how much detail per factor or factor interaction) contributes to the expense of building and operating the model. As a result, modeling and simulation tend to be an expensive option that needs to provide a rationale for its cost-benefit ratio. If the benefit, measured in terms of lives or money saved through improved decisions, compares favorably to the resources required to build and operate the model, then the model is considered useful and worthwhile.

9.4.3 Effects Prediction Tools for Planning

As mentioned in Section 9.1.4, when in a relatively unknown or poorly understood situation, the first concern of the planner is how to visualize the situation and option space [2]. Simply being able to visualize the situation and its inherent interactions is beneficial to the planner and commander.

In addition to visualizing the current situation, the commander and planners will want to know what is likely to happen in response to given actions. There are a number of commercially available, albeit general, prediction methodologies are available. Extend, Ithink, and Analyst's Notebook are three COTS tools that have sometimes been adapted for use in EBP. Extend is a process or workflow model, Ithink is a systems dynamics modeling tool, and Analyst's Notebook is primarily a relationship display tool (albeit sometimes used to "predict" previously undiscovered relationships). In addition, the U.S. Air Force's Predictive Battlespace Awareness project was an effort to bring together a wide range of tools to support planning that projected anticipated outcomes using numerical computation [18].

The Java Causal Analysis Tool (JCAT) is a government-developed tool written in Java that allows the user to define and evaluate a Bayesian Belief Net representing the system being investigated [19]. The Operational Assessment Tool Bayesian Belief Net capability in SDT described earlier was a variant of this JCAT tool.

Markov and Hidden Markov Models (HMMs) have also been used to predict organizational relationships currently unseen by the outside observer [20]. In HMMs, the observed evidence of an unknown organization's behavior is compared to templates of sample organizations to see which ones are mostly likely candidates for the one being observed [21].

Exploratory modeling has been used to hypothesize a wide range of outcomes to better understand the response surface of the system. The intent of exploratory modeling is not so much to predict a single outcome, but to predict a range of outcomes given a set of input conditions so the user has a better understanding of the tradeoffs involved in interacting with this system [22].

An example of applying the principles of exploratory analysis has been undertaken in the Integrated Battle Command (IBC) project described briefly earlier. The intent of the IBC project models, and in prediction models in general, is not to provide the right answer as a point solution, but to help define the space of plausible outcomes, including both good opportunities and unexpected disastrous outcomes. Understanding the response surface of the models helps planners modify their plans to help prevent or mitigate possible bad outcomes and better identify and exploit possible good outcomes.

In order for the IBC models to predict a range of outcomes, the normal interactions among all of the PMESII elements must first be understood at least to some degree of accuracy, and then the effects of DIME actions taken in that environment, interacting with all of PMESII factors. To make matters even more challenging, the desired effects are not just disruptive, as would be in the case of war, but may often need to be *positive* when involving stabilization and nation-building operations. Since the intended IBC model suite is trying to model some of the most basic DIME and PMESII interactions, the subject of these models is far beyond traditional military operations. These models are applicable to interagency planning, including the State Department, Commerce, Intelligence, and others, as well as military planning. Such a model suite would also help define the role and contribution of information operations in interagency planning, which is also the subject of Chapter 12. Assuming the IBC project succeeds, it will help provide higher-level guidance and context from the interagency level to the military commanders and the IO planners. Since the IBC project is only in its first year, however, actual fielding of any useful components may take some time.

Planners do not simply want to be able to predict the outcomes of their own actions, however. They also want to predict the outcomes of enemy decision processes and actions. This brings us to the subject of Section 9.5.

9.5 Adversarial and Nonadversarial Reasoning Technologies

Adversarial and nonadversarial reasoning technologies can be considered a subset of effects and prediction technologies, in that the technologies are trying to both describe and predict how adversaries, neutrals, and groups of shifting allegiances perform their planning and decision-making. Section 9.5.1 assumes that the adversary is highly unlikely to switch allegiances during the planning horizon, whereas Section 9.5.2 addresses how to plan for adversaries, neutrals, and allies that might switch allegiance in combat situations. Section 9.5.3 discusses the same issues, only for noncombat situations. While still in their infancy, some promising efforts are underway to represent adversarial and nonadversarial reasoning to provide insights to friendly planners.

9.5.1 Adversarial Reasoning

As mentioned in Chapters 1 and 3, how the enemy reacts to the information received can have a significant effect on the success or failure of an information operation. Enemy commanders base their decisions on the information they have available and the level of confidence that they have in that information. Enemy operators of communications and information systems base their actions and reactions on the information they have available. If our (friendly side) planners can model or otherwise represent the reasoning processes of our opponents, then we would have a better chance of predicting their reactions to the information they observe (or expect and do not observe), thereby increasing the probability of success of the friendly information operation.

The types of questions an adversarial reasoning tool might want to answer include:

- What is the decision-making structure (e.g., hierarchy, cells, and so forth) of the adversary?
- Who makes which decisions and how often?
- Who provides inputs to the adversary's decision process?
- Can we predict with any reasonable probability the decisions the adversary will make given a set of known or assumed access to a certain set of information?
- What information can we provide or deny the adversary to help make their decisions favorable to our objectives, or at least less unfavorable to our objectives?
- How do we measure the success of our adversarial reasoning models if the reasoning process is unknown and only the final decisions are known?

These are important considerations in IO, and ones planners would like to have some insight into when planning IO missions. However, understanding the inner workings of an enemy's reasoning and decision-making processes is not an easy task. Without an inside agent to let you know exactly what is happening, that reasoning process must be inferred from indirect evidence.

Like most intelligence processes, this inferential evidence can be described in terms of externals and internals. Inferences based on externals are accomplished by tracking things like who is talking to whom without understanding the content of those communications. Inferences based on internals are achieved by understanding the content of those communications. Although one would prefer to have access to both, one can often only obtain access to one, and more often than not, only the externals.

Still, a lot can be learned by monitoring the externals. Just knowing who is talking to whom at certain times can indicate that decisions of a certain magnitude are being discussed. When the press reports that the U.S. intelligence indicates an increase in the chatter on terrorism channels, it often indicates something is up. This is similar to intelligence collected by reconnaissance by fire in the Vietnam War. Even if the enemy did not fire back, but the enemy's radio traffic suddenly increased, it usually meant you were shooting near something of importance and should probably continue that activity. So it is, too, with monitoring externals for adversarial reasoning.

When internal contents are available, more information is available to build the "model" of the adversarial reasoning process. Note that when we say "model," this is not necessarily a detailed simulation of what the enemy is thinking. On the contrary, there is a heavy emphasis on probabilities and bounding possibilities due to the high level of uncertainty in such an inference problem. To make matters more complicated, there is usually not just one model, but many models, each providing a context to help extract inferences out of any given observable.

The number of prestored inference relationships that can be checked very quickly is enhanced by computers. Once the inferential or evidential knowledge has been entered into the computer, the computer can quickly search through those rules, trigger the ones that apply, and repeat the process up the chain of inferential or evidential reasoning until it spits out a set of probabilities for what it thinks is happening and how the enemy is reasoning right now. As with all computer programs, garbage in means garbage out, so if the inferential rules are flawed, then the inferences are useless. However, these rule sets can be tested, and when they work, they could be expected to provide real value added to the friendly planner.

These inference rules are usually arranged in a hierarchical fashion, so that lower-level inferences can be built upon by the next higher inference rule sets. For example, just observing who is talking to whom based on externals might lead to inferences about who reports to whom in the organizational structure. Adding in some basic content, such as whether these messages are questions, can further infer relationships in a hierarchical organization. Further content analysis might be able to infer the organizational structure of the group of interest with a substantial degree of accuracy.

Note that when inferences build on inferences, there is the possibility of leading oneself out on an evidential limb of reasoning that is entirely logical but may be completely false. Even so, the adversarial reasoning models seem to do fairly well at predicting basic organizational relationships.

In order for data to be quickly digested into any predictive computer model, it must be in a format the computer understands. Since the raw data is usually fairly voluminous, reformatting it by hand is not an option. However, if one were able to take raw data feeds and process them directly, one could have a useful tool. One such example was developed by Kathleen Carley at Carnegie Mellon University, who took raw e-mail files from a sample scenario and fed them directly into a program that analyzed the header information for who was talking to whom and when and then developed a fairly accurate node-relationship display using social network analysis tools [23]. The ability to process the raw data directly is essential to developing an adversarial reasoning model that will work in the real world.

A number of adversarial reasoning approaches and models have been developed over the past few years. For example, in 2002 Bell, Santos, and Brown presented an approach for adversary decision modeling using intent inference and information fusion. They described a three-tiered approach in which raw data would be collected by intelligent mobile agents, information fusion technologies would provide higher-level evidence based on their rules and algorithms, and an intent inference engine would model the interests, preferences, and context of the opponent in order to suggest likely enemy intent [24].

Another example is the subsequent work by Surman, Hillman, and Santos, which attempted to enhance the war-gaming of Blue versus Red COAs by using a model of Red's reasoning to generate their own COAs based on an inference engine based on Bayesian Belief Nets. Their intent was to sufficiently represent Red reasoning to generate emergent behavior in the wargame and thereby predict unexpected enemy courses of action [25].

The SDT model described previously is another example of an adversarial reasoning model. SDT attempts to assist the planner and warfighter in the development and assessment of plans that create a change in the opponent's behaviors. SDT follows the following four planning and execution steps:

1. Desired behavior in opponent;
2. Friendly operations and anticipated responses;
3. Outcomes of actual operations;
4. Actual responses by the opponents.

McCrabb and Caroli noted that it is difficult to incorporate behavioral models in the current COA war-gaming schema, and that major technical challenges will have to be overcome to solve this problem [11]. The SDT proponents also expressed the concern that predicting and then assessing how physical actions spawn behavioral effects resulting from our actions is itself *the* major challenge. This is why we described in Chapter 6 the need to define the measure of causal linkage to explicitly identify the assumptions required to connect actions to desired effects.

At the time of this writing, the state of the art of adversarial reasoning technologies is still in its infancy, and even the most promising appear to be most likely to succeed at the higher echelons, where more time is available to analyze and understand the situation and predict outcomes. Tactical-level adversarial reasoning technologies are currently hampered by the lack of raw data that can be accessed and processed quickly enough to provide timely, useful predictions.

9.5.2 Nonadversarial Reasoning: Shifting Allegiances in Combat Situations

Nonadversarial reasoning is similar to adversarial reasoning, except that the intent is to characterize, model, and predict the decision processes and actions of groups that are not necessarily permanent adversaries or allies. Nonadversarial reasoning applies to all groups that are nonadversarial players, as well as to all groups currently on *any* side that may follow shifting allegiances. This section discusses handling the shifting allegiances and bribes in combat situations, while Section 9.5.3 describes handling shifting allegiances in noncombat situations.

When it is possible for a group, or members of a group, to change allegiances during the course of an operation, that can have a significant effect on both the planning and the successful execution of a plan. If one assumes that all players involved in the upcoming plan, both enemy and friendly, are rigidly members of one side or another, then our planners and warfighters can be blindsided by the fact that some of our allies may defect or be bribed into cooperating with the enemy. For example, when assaulting Fallujah the first time, much of the Iraqi Fallujah Brigade folded, with a number of soldiers subsequently defecting to the insurgents and terrorists [26]. In a similar manner, it was difficult for Coalition commanders to fully consider the likelihood of how many of our Afghan allies would accept bribes from al-Qaeda or the Taliban, or that the outcomes would be that bin Laden and hundreds of his fighters would be able to escape from Tora Bora [27].

U.S. planners must improve their ability to include shifting allegiances and outright bribery as part of standard operating procedures. Throughout history, there have been a number of key defections from one side to the other and cases in which bribes succeeded in creating great harm to one side, to the advantage of the other. The defection of Governor Khayrbak to the Ottomans during the Battle of Marj Dabiq in 1516 ended Mameluke rule in Egypt [28]. The defection of Kobayakawa to Tokugawa during the battle of Sekigahara sealed the Shogunate for the latter in 1600 [29]. In twelfth- and fifteenth-century England, nobles switched sides, sometimes more than once, in the struggles between Stephen and Maude [30] and the War of the Roses, respectively [31].

While it is true that these examples of switching allegiances are associated with feudal or tribal situations, many of the groups we are currently dealing with in the modern world still consider themselves tribal, and some of their relationships are often based on feudal principles. In many cases, the concept of nationalism is, at best, a distant second over tribal allegiances.

In some of these historical cases, the switch by a leader from one side to another was very much in the balance, depending on which side appeared to be winning. The side that relies on a member of the opposing force to defect may have his plans severely constrained until the defection takes place. For example, at Sekigahara, the last great field battle for the Japanese Shogunate in 1600, Kobayakawa Hideaki defected much later in the battle than Tokugawa Ieyasu (the eventual winner) would have liked because Kobayakawa was not certain which side would win. Only after Tokugawa sent arquebusiers to fire at Kobayakawa's troops with harassing fire and sent a subordinate chieftain to threaten his personal advisor did Kobayakawa finally decide to change sides [29].

Bribes were often used by the Chechens to pass through Russian lines. Even so, not every bribe event went as planned. For example, when the Chechens were trying to escape from Grozny in February 2000, they bribed a Russian officer to let them pass through his area. However, the path provided by the Russians led through a minefield, and some reports described artillery further adding to the devastation. While some still escaped, their losses were horrendous and key leaders were killed while others were seriously injured [32, 33].

The first step, therefore, in preparing our planners for the potential problems and opportunities provided by shifting allegiances is to *make sure they understand that such shifts in allegiance or bribes are possible.* To help ingrain this training objective, it is important to include at least one group with shifting allegiances on each side during training exercises just to get planners used to the concept and possibilities for each side. As long as the planners are warned ahead of time that such shifting allegiances are possible for either side, they will generally adapt by planning options and COA development that account for those possibilities.

The second step is for our planning and execution monitoring tools to be capable of handling shifting allegiances on the battlefield. For example, the Future Combat System is intended to help identify which detected assets are enemy and which are friendly. But what if the allegiance of a number of these assets changed during the course of an operation? Could our tools and displays handle that real-time side change? Conversely, if we are using Blue Force Tracking to identify friends, and a group that started out on our side suddenly turns against us, will our tools be capable of identifying the now-opposing forces as enemies? These are not trivial problems, both from the system design perspective and from a warfighter perspective. While not exactly an information operations planning issue, it is definitely a potential problem with respect to the information upon which friendly decisions will be made.

For the planner, there are five important shifting allegiance considerations when developing COAs and selecting plans:

1. Whether a defection or bribe of a *friendly* group or unit by the enemy is possible and, if so, did we pay them more than the enemy did;
2. Whether a defection or bribe of an *enemy* group or unit is possible;
3. Whether and how much the defection or bribe of an enemy group can be relied upon in the plan, and how to plan contingencies if the event does not occur as planned;
4. How to plan friendly operations with contingencies to handle the possible unexpected defection or bribe of friendly groups during an operation;
5. Whether an intended defection or attempted bribe of a friendly group by an enemy can be detected and leveraged for increased success.

Note that information operations, and in particular influence operations, are primary mechanisms for determining which enemy groups may defect or be bribed, and whether these groups may be relied upon to follow through with their promise. There is, however, little military science of defections and bribes within U.S. military doctrine. We have not traditionally been trained to think that way, yet many of our allies and opponents live and operate in environments in which allegiances shift back and forth over time.[9]

To help overcome the lack of experience in this area, the U.S. military can use two techniques. The first is to leverage the cultural knowledge of our coalition partners who are more familiar with environments of shifting allegiances. The second is to develop nonadversarial reasoning tools to assist the planner in the assessment of such options. This is the next step beyond simply

---

9. While it is true that in World War II the Allies tried to persuade the Vichy French forces in North Africa not to oppose the U.S. landing, these negotiations were without success until many days after the landing, the U.S. negotiators were winging it, and at least one negotiator was killed [34].

understanding that defections and bribes are possible and will hopefully provide some probability and confidence as to whether defections or bribes are likely to occur, whether they will appear to succeed, and whether they will actually succeed. Some of the factors to consider are in such a model are:

- Are members of the two opposing sides related by tribe, blood, marriage, or religion?
- Is one of the groups or group leaders recently disaffected? (For example, Kobayakawa was publicly berated by his leaders shortly before the climactic battle [29], and Benedict Arnold was the subject of two Continental Congress investigations [35].)
- Was one of the groups or group leaders forced into cooperating with our opponents in the first place (like the Vichy French), or was one of our groups coerced into cooperating with our side?
- Which side can offer the most personal gain to each group? (For example, one of the groups bribed at Tora Bora was paid by the Coalition, but it had already been paid more by al-Qaeda [27].)

Getting this type of information and other information required by IO planners is described in Chapter 10. In many cases the best data is not directly available, and what is known must be inferred from indirect sources. The technical details of nonadversarial reasoning models are very similar to those of adversarial reasoning models, in that there will be a series of rules providing context so that evidence is compared to the knowledge base and inferences are generated. Again, the main problem is whether the nonadversarial reasoning tool will take the planner out on an evidential limb, which is why the planner should also include contingencies (branches and sequels) for whether the defection or bribe occurs at all, and whether it occurs when and as planned.

9.5.3 Nonadversarial Reasoning: Shifting Allegiances in Noncombat Situations

Section 9.5.2 described the need for and benefit of nonadversarial reasoning tools in combat situations. This section will address a similar concern in noncombat situations. However, nonadversarial reasoning for noncombat situations goes far beyond assessing the probability that a group might be bribed or defect during combat. It involves attempting to understand the long-term interests and options available to all the key players in a given region. This type of nonadversarial reasoning is of significant interest to those involved in planning peacekeeping operations, stability operations, and nation-building.

Our planning tools must be designed to be more than two-sided. It is not just us versus them. That mindset will only result in a self-fulfilling prophecy. If we label an enemy as once an enemy, always an enemy, he will remain an enemy. But if our decision and planning tools account for the fact that the allegiances of our current enemies may change over time, then we have a better chance of planning for a future where the number of enemies is small (although it will is unlikely to ever be zero).

In a similar manner, our planning tools need to account for the fact that our current allies may not remain our allies. Some of our NATO allies have openly declared themselves to be in opposition to the United States simply because the United States is the sole remaining world power [36]. Our planning must take into account the fact that today's friends might be tomorrow's enemies, and vice versa.

Of particular interest to noncombat nonadversarial reasoning tools is a characterization of the interests of each group and leader that evolves dynamically over time. What are the goals of each group or leader? What are their capabilities and current intent? Can you provide alternatives to them that they may not yet have considered and that can help them reach or advance toward their goals?

One of the most successful nation-building operations has taken place in Afghanistan. The Taliban has not been able to generate sufficient public support to create an insurgency as is occurring in Iraq, and the central government in Afghanistan appears to be gaining increasing control over time.

Part of the reason for the success of this program has been the way the various groups and leaders within Afghanistan have been encouraged to cooperate with the National Afghani Government, as well as with NATO and Coalition forces. One of the lessons learned from the Afghanistan stabilization effort is that there frequently are not any clear "good guys" (at least by U.S. standards), and that the "bad guys" don't always remain "bad." To quote Dr. Sean Maloney [37], "… it has proven to be far better to assume everybody is 'dirty' after 25 years of war and to start anew."

Based on the principle of working with what you have, most of the recalcitrant players in Afghanistan have been co-opted into the central government structure, given a promotion to a government position with a broad portfolio of responsibility and authority, and provided with experienced staff to help mentor them in their new duties. Meanwhile, their "second-tier" militia leaders were promoted to become police commanders—but in another province, with other forces funded by Kabul [37].

Working in parallel, some militia units were disbanded under the Disarmament, Demobilization, and Reintegration (DDR) program, so that when the Afghan National Army entered the area, the militias were too weak to stop them [37]: "To a certain extent, law and order remains relative, but the concept behind an incremental transfer of power applies. The method of establishing a small Afghan National Army garrison, building it up slowly, and having its personnel develop relationships with militia forces provides yet another mechanism for progress."

These cooperative techniques are based on attempting to find a win-win situation for recalcitrant players so that they are brought into a power-sharing arrangement, rather than alienated and carrying on the fight. Sun Tzu said [38], "Keep your friends close and your enemies closer." To a great degree, the new paradigm is now: "Keep your enemy close until he is no longer your enemy."

For example, the Sunni insurgents in Iraq, many of whom had vowed to fight the Coalition until the bitter end, have realized that armed radical Shi'ite groups, Iranian influence, and international terrorists are greater threats to their long-term survival than the United States and the other Coalition members. As a result, they have begun negotiations to identify areas of common interest [39]. The Sunni Iraqis in particular have been receptive to the admission by the U.S. forces that mistakes have been made, and that there is an opportunity to share the blame as part of the effort to come to the negotiating table [40].

The intent of nonadversarial reasoning models for noncombat situations is to represent how different groups reason. As long as the groups represented include a wide range of the spectrum of allegiances and it is understood that these allegiances can change over time, the principles and approaches of adversarial reasoning should apply. Due to the complexity of the issues and their interactions, it will be some time before these models are sufficiently mature to effectively support real-world planning. However, there are opportunities to start with very simple visualization tools to help identify the various groups' positions and how they may evolve over time.

For example, one of the most needed tools is one that helps planners identify areas of common interest, or at least areas of interest that most parties can live with. Figure 9.3 was provided as one useful display in the Integrated Battle Command Project, Phase I. While it is a simple visualization tool, it can help get people discussing common interests and fears regarding complex political topics.

The horizontal line of Figure 9.3 shows a hypothetical spectrum of possible end states related to the issue of Kurdish State Autonomy. At the left end is the formation of a Kurdish nation, while at the right end is no separate Kurdish identity. Above the line are a set of group names and arrows. The arrows represent the direction in which the specific group is likely to try and influence the possible end states. Turkey and Iran, for example, do not want to see an independent Kurdish State because they are concerned that their own indigenous Kurdish populations will seek the same. The Iraqi Sunnis, and to some degree the Iraqi Shi'ites, do not support the creation of a Kurdish federal state for fear that state might later separate from Iraq. The Kurds, of course, seek as much autonomy as possible, because they fear losing the identity and relative autonomy that they have recently gained. The United States would like to see a

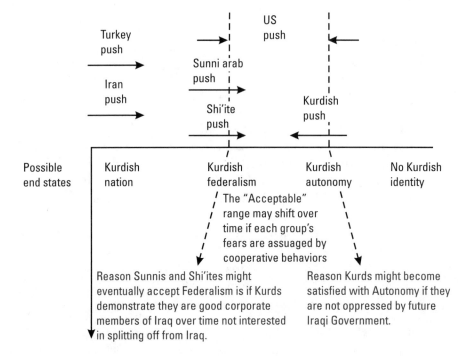

Figure 9.3 Representation of competing interests and areas of acceptance.

feasible solution and therefore is trying to encourage all players to enter into a potentially acceptable range of possible outcomes.

Below the horizontal line, the acceptable range boundary expands outward over time (where time goes forward down the figure). This reflects the fact that most of the push provided by the various groups is more a matter of their fears (founded or otherwise) that any move in a given direction will end up one extreme end or the other of possible outcomes. Although this is a possibility, as long as the groups with competing interests remain fairly balanced, the outcomes are likely to remain in the acceptable range. However, the view of what is acceptable depends upon what has occurred recently to allay fears. If the new Iraqi government is equitable to the Kurds and the Kurds are satisfied with their current level of autonomy, then their push toward a Kurdish state, or even a federal state, will be less intense. Conversely, if the Kurds continue to demonstrate that they are demonstratably corporate citizens of a unified Iraq, then Iraqi fears that Kurdish federalism will inevitably lead to an independent Kurdish nation will subside over time.

Of course, real issues are often more complex than the situation described above, and not every issue's possible end states will lend themselves to a one-dimensional spectrum arrangement as shown above. But this type of display does demonstrate the type of relatively simple tool that can be used to help visualize the competing interests of all parties involved on a given topic.

Overall, planning tools for noncombat nonadversarial reasoning tools are, again, very similar in their approach, design, and limitations to the adversarial reasoning tools. Solid data is used when available, inferential data is derived from indirect information, and based on the quality of the knowledge captured in the context rules in the tool, useful insights about nonadversarial reasoning in noncombat situations could be provided to the planners and warfighters. Such tools are still very much in their infancy, but since the future appears to include more stabilization and nation-building operations, such tools will be in high demand [41].

In summary, the IO planning tool environment has made great strides in the last few years. The planning processes and workflows now have semiautomated planning tools that help increase the efficiency of the planning staff, and thereby reduce the OODA cycle time. However, there is always room for improvement. Information technology is changing rapidly. The global information space in which the U.S. military must operate is quickly evolving. For planning and other decision support tools to survive, flexibility and scalability must be built in. Moreover, the increasing availability of significantly enhanced real-time information feeds will be essential to the success of future IO planning and execution. In particular, real time feeds and their quality assurance, as well as the necessary bandwidth and multilevel security, will be required. And most importantly, any such planning tools will need to be fully integrated as one part in an overarching system that includes all aspects of military operations: doctrine, organization, training, materiel, leadership development, personnel, and facilities, which we will discuss further in Chapter 10.

References

[1] DoDD 3600.1, Information Operations Policy (December), Washington, D.C.: Government Printing Office, December 2003.

[2] Allen, Patrick D. "Visualization Tools: First Choice for the Planner," *Phalanx*, September 1998.

[3] Allen, John G., "Decision Support Tools for Effects Based Campaign Planning and Management," DARPA Integrated Battle Command Briefing, August 2005.

[4] JFSC Pub 1, *Joint Staff Officer's Guide 2000*, National Defense University, Joint Forces Staff College, 2000.

[5] Field Manual FM 5-0, *Army Planning and Orders Production*, Headquarters, Department of the Army, January 2005.

[6] Wentz, Larry, and Lee Wagenhals, "Integration of Information Operations into Effects-Based Operations: Some Observations," the Command and Control Research Program (CCRP) Web site, 2003, http://www.dodccrp.org/events/2003/8th_ICCRTS/pdf/072.pdf.

[7] Pioch, Nicholas J., *Continuous Strategy Development for Effects-Based Operations*, Final Technical Report, February 2001–November 2005, BAE Systems, Burlington MA.

[8] Wilkins, David, and Marie DesJarins, "A Call for Knowledge-Based Planning," *AI Magazine*, Spring 2001.

[9] AFDD 2-5, *Information Operations*, Air Force Doctrine Document 2-5 Draft, 2003

[10] Henningsen, Dr. Jacqueline R., et al., *Analyzing Effects Based Operations Workshop Report*, Military Operations Research Society, January 29–31, 2002.

[11] McCrabb, Dr. Maris "Buster," and Joseph A. Caroli, "Behavioral Modeling and Wargaming for Effects-Based Operations," *Military Operations Research Society Workshop on Analyzing Effects Based Operations*, January 29–31, 2002, http://www.mors.org/meetings/ebo/ebo_reads/McCrabb_Caroli.pdf.

[12] White, Christopher, Strategy Development Tool (SDT) Overview, BAE Systems, briefing, July 2006.

[13] Byrne, John, "Background Paper on System Dynamics as a Tool for Effects-Based Operations (EBO) Analysis," *Military Operations Research Society Workshop on Analyzing Effects Based Operations*, January 29–31, 2002, http://www.mors.org/meetings/ebo/ebo_reads/Byrne.PDF.

[14] Robbins, Matthew JD, *Investigating the Complexities of Nationbuilding: A Sub-National Regional Perspective*, thesis, Air Force Institute of Technology, Wright-Patterson Air Force Base, Ohio, March 2005.

[15] Colaizzi, Jennifer, "Experimentation Director Looks Back on Tour, Reflects on What's Next for Directorate," U.S. Joint Forces Command press release, October 2, 2004.

[16] Cupp, Sgt. Jon, "USJFCOM Teams with Purdue University to Add the Human Factor to War Game Simulations," U.S. Joint Forces Command press release, February 6, 2004.

[17] Allen, Patrick D., Jim Gillogly, and Jim Dewar, "Non-Monotonic Effects in Models with Stochastic Thresholds," *Phalanx*, December 1993.

[18] Snodgrass, Brig. Gen. Michael, "Effects Based Operations," briefing for C4ISR Summit at Danvers, MA, August 20, 2003.

[19] Lemmer, J., "AFRL Demonstrates Java Causal Analysis Tool for CENTCOM," *AFRL Monthly Accomplish Report Executive Summary*, September 2005, http://www.afrl.af.mil/accomprpt/sep05/accompsep05.asp.

[20] "Hidden Markov Models," http://en.wikipedia.org/wiki/Hidden_Markov_model, May 2006.

[21] Singh, Satnam, et al., "Stochastic Modeling of a Terrorist Event Via the ASAM System," *IEEE Conference on Systems, Man and Cybernetics*, The Hague, the Netherlands, October 2004, http://www.engr.uconn.edu/~sas03013/docs/SMC_2004_Stochastic_modelling_via_ASAM.pdf.

[22] Bankes, Steven C., and James Gillogly, *Exploratory Modeling: Search Through Spaces of Computational Experiments*, RAND Report RP-345, 1994; http://www.rand.org/isg/modeling.html.

[23] Carley, Kathleen M., "Social and Organizational Networks," Carnegie Mellon Web site, Center for Computational Analysis of Social and Organizational Systems (CASOS), May 2003, http://www.casos.cs.cmu.edu.

[24] Bell, Benjamin, Eugene Santos, and Scott Brown, "Making Adversary Decision Modeling Tractable with Intent Inference and Information Fusion," *Proceedings of the 11th Conference on Computer Generated Forces and Behavioral Representation*, Orlando, FL, May 2002.

[25] Surman, Joshua, Robert Hillman, and Eugene Santos, "Adversarial Inferencing for Generating Dynamic Adversary Behavior," *Enabling Technology for Simulation Science VII Conference*, Orlando, FL, April 2003.

[26] "U.S. Tries Again with Iraqi Force, This Time Emphasizing Consensus," World Tribune.com, May 13, 2004, http://216.26.163.62/2004/ss_iraq_05_12.html.

[27] Smucker, Philip, "How Bin Laden Got Away: A Day-by-Day Account of How Osama Bin Laden Eluded the World's Most Powerful Military Machine," *The Christian Science Monitor*, March 4, 2002, http://www.csmonitor.com/2002/0304/p01s03-wosc.html.

[28] Rashba, Gary L., "The Battle of Marj Dabiq Saw a Change in Warfare and Rulers in the Middle East," *Military History*, April 2006.

[29] Murphy, John F. Jr., "Battle of Sekigahara: Shogun's Rise to Power," *Military History*, May 2005.

[30] Ashley, Mike, *British Kings & Queens*, New York: Barnes and Noble Books, 2000, pp. 512–516.

[31] Knox, Dr. E. L. Skip, "The War of the Roses, Consequences of the War," Europe in the Late Middle Ages course, History 309, Boise State University, undated, http://history.boisestate.edy/hy309/wotr/23.html.

[32] Hunter, Chris, "Chechnya," *Center for Peace Keeping and Community Development Journal*, 4.1, undated, http://maic.jmu.edu/Journal/4.1/chechnya.htm.

[33] "Chechens in Minefield Nightmare," *The Irish Examiner online*, February 5, 2000; http://archives.tcm.ie/irishexaminer/2000/02/05/current/fpage_2.htm.

[34] Wade, Billy, "The Spark of the Torch," *The Dispatch* magazine, Volume 17, Number 3, Fall 1992, http://rwebs.net/dispatch/output.asp?ArticleID=44

[35] "Benedict Arnold: Revolutionary War Major General," Benedict Arnold Web site, Evisum Inc., 2000, http://www.benedictarnold.org/.

[36] Vinocur, John, "France Has a Hard Sell to Rein in U.S. Power," *International Herald Tribune*, February 6, 1999, http://www.iht.com/articles/1999/02/06/france.t_1.php.

[37] Maloney, Sean M., "Afghanistan Four Years On: An Assessment," *Parameters*, U.S. Army War College, Autumn 2005, http://www.carlisle.army.mil/usawc/Parameters/05autumn/contents.htm.

[38] Tzu, Sun, "Sun Tzu Quotes," http://www.quotationspage.com/quote/36994.html, May 2006.

[39] Johnson, Scott, Rod Nordland, and Ranya Kadri, "Exclusive: Direct Talks—U.S. Officials and Iraqi Insurgents," *Newsweek,* reported on MSNBC online news, February 6, 2006, http://www.msnbc.msn.com/id/11079548/site/newsweek/.

[40] Ricks, Thomas E., "U.S. Unit Masters Art of Counterinsurgency," *The Washington Post*, http://www.msnbc.msn.com/id/11376551/, February 16, 2006.

[41] Barnett, Thomas P, M., *The Pentagon's New Map*, New York: G. P. Putnam's Sons, 2004.

10

Planning for IO Logistics, Training, and Acquisition

> *Amateurs talk about strategy; Professionals talk about logistics.*
> —General (Ret.) R.H. Barrows, Commandant, U.S. Marine Corps

> *I don't know what the hell this logistics is that Marshall is always talking about, but I want some of it.*
> —Fleet Admiral E. J. King, 1942

This chapter discusses the broad areas of planning related to the logistics of information operations. In order for IO to work, a lot of background preparation must be done across all supporting aspects of IO, including doctrine, organization, training, materiel, leadership development, policy, and facilities (DOTMLPF). Each one of these supporting areas requires planning and the execution monitoring of those plans. Note that the DoD-defined categories of IO that contribute directly to the logistics of IO are electronic warfare support (ES), counterintelligence, and combat camera, as well as intelligence capabilities in general (see Figure 4.1).

Section 10.1 discusses the supply-and-demand aspects of IO. As mentioned in Chapter 3, IO planning has a nearly insatiable appetite for a wide range of information. Those that plan IO require a great deal of very specific information from the information suppliers—the intelligence community—who together form the supply chain for IO information support. Due to the complexity of the Information Age, many intelligence agencies may be required to provide the necessary information to prepare an information operation, and many organizations that are preparing information operations may use the same information provided by one intelligence agency. In order to plan and execute IO in a timely and effective manner, the necessary information must be readily available, which is measured by *information readiness*.

Section 10.2 presents issues of transformation. The whole U.S. military and other interagency staffs are involved in transforming to operate better in the Information Age, especially with the new threat of non-nation-state actors. After discussing the DOTMLPF issues, this section further examines the IO acquisition process, and the lack of mature tools to support IO acquisition planning.

Section 10.3 focuses on the key events in three components of DOTMLPF: training exercises, transformation experiments, and materiel acquisition testing. Each of these three types of events has unique challenges when trying to include IO, and this section describes ways to overcome these challenges.

Section 10.4 discusses selected legal aspects of IO. What constitutes use of force, and therefore a belligerent act, is still not well defined in IO and will probably continue to evolve. At the same time, there are a number of assumptions about the nature of IO in conflict that need to be discussed. This section also presents how to help planners and commanders know what is legally allowed both at the start of and throughout the planning and execution monitoring process. Lastly, the complex issue of Title 10 (military) versus Title 50 (intelligence) overlapping areas of responsibility is discussed at the end of this chapter.

10.1 The Logistics of IO

Offensive, influence, and defensive IO all require large quantities of very detailed and up-to-date information. This section discusses why and how the planners and commanders can help rein in and prioritize their information requirements so that they are actually achievable. Section 10.1.2 describes the organizations that supply the information, and how the DoD is attempting to improve this process. Section 10.1.3 discusses the concept of information readiness, which is the process of having the information you need to accomplish the mission prepared and available.

10.1.1 Information Requirements for IO Planning

One of the big problems of IO planning is that it has a nearly insatiable appetite for information to support the achievement of IO missions. To plan for disrupting an enemy network, the planner needs information about the target network and its operations. To plan on defending a friendly network, a planner needs timely information about our own networks. To plan an influence operation, information is required about the groups and messages in the idea battlespace and their interactions.

Theoretically, if the planner knew everything about the target system or group, one could plan the perfect operation and have high confidence that the plan would execute flawlessly. As a result, early requests to the intelligence community were to provide all possible information about every target—a request that the intelligence community correctly ignored. Intelligence resources are scarce compared to the demands for information placed on them, and the lack of focus in these early comprehensive requests for information to support IO caused them to be assigned the lowest priority.

This section will describe some of the types of information requirements necessary to support IO planning. Here we define the demand for information, while Section 10.1.2 will discuss sources of information supply. We also describe some differences between offensive, influence, and defensive IO information requirements.

On the offensive side, IO planners would like to know everything about potential target networks. The following list shows some of the types of information required to support IO targeting:

- A description of the enemy in general;
- A characterization of each target;
- Vulnerability analysis of the target;
- Susceptibility analysis of the target to different attack options;
- COA analysis to compare alternatives and employment options;
- Synchronization analysis to synchronize and deconflict effects, including spectrum managment of electronic warfare;
- Campaign analysis, if required, at higher echelons;
- Plan selection analysis, or how to pick the best plan.

Note that different types of information are required to support each step in the offensive planning process. Gaps in the required information do not stop the targeting process, but do increase the risk of something going wrong. Military commanders are trained to make decisions in the absence of information. As a result, the job of the information suppliers is to improve the decision that would normally be made in the absence of sufficient information. Like planning traditional military operations, the key to solving the problem in IO is how to plan with sufficient regard to what is currently known and what can be obtained in time to support a desired operation.

For influence operations, there is often scant objective data with which to support planning. Moreover, the target set is constantly changing. Descriptions of groups and their interactions with each other constantly shift, and the understanding of such groups and interactions changes even more often. This inherent complexity of influence operations means that we will (more often than not) require human judgment and subject matter expertise to assist in influence operation planning, analysis, and assessment. The intricacies of different groups and their interactions with each other and with messages in the idea space are also often subtle and difficult to measure. Even so, modern survey techniques, algorithms, and data mining techniques should be useful in helping fill in the gaps in the unfilled information requirements, and could provide a mechanism to help keep the complexity manageable.

On the IO defense side, the Joint Task Force for Computer Network Defense (JTF-CND) is tasked [1]: "to coordinate and direct the defense of DoD computer networks and systems." To do so, they need accurate and timely data about the information networks of the various Services. However, the Services have no incentive to share the data about their networks with the DoD or Joint Offices. As a result, JTF-CND is sometimes attempting to manage and protect a network about which it has limited or dated information.

The solution to this problem is to realize that no one will ever have as much information as he desires to perform IO with a high degree of confidence. Information requirements or requests need to be as specific as possible about the information required, why it is needed, and why it is needed in a specified time frame. To help ensure that the correct information is provided in a timely manner, requests for information should include as much context information as possible, such as why the information is needed, in case alternative or related information is available.

Note that there are three general categories of data:

- Hard, quantitative objective data based on measurable and repeatable outcomes, such as those contained in the *Joint Munitions Effectiveness Manual;*
- Quantified subjective data, where numerical values are assigned subjectively to define data, such as a ranking scheme, a scoring mechanism, or other numerical representation of subjective information;
- Soft, subjective data, which are basically opinions (sometimes informed), but often the only type of data available.

These three categories of data are listed in order of most to least desirable for the decision-maker. They are also listed in the order of least to most likely to be available. When the IO JMEM is finally created, it will help make more hard, quantitative data available for planners and commanders, but that may be a number of years in the future.

10.1.2 Information Suppliers

There are many information suppliers, including open-source information, such as online maps and satellite imagery. For the most part, U.S. military planning relies on the intelligence community (both within and outside of the Defense Department) to provide the information required to support IO.

The information suppliers work full-time collecting the information that planners and warfighters might need when they undertake missions. Since no one can know everything about everything, the intelligence agencies focus on areas of interest as requested by contingency planners in peacetime and focus even more on quick-response requests for information during conflicts.

Due to the complex nature of today's world, different intelligence agencies have tended to focus on different areas of expertise. For example, the National Security Agency (NSA) mission statement from their home page reads [2]: "The ability to understand the secret communications of our foreign adversaries while protecting our own communications … gives our nation a unique advantage." The Defense Intelligence Agency (DIA) Web site states that [3]: "We cover all aspects of military intelligence requirements—from highly complex missile trajectory data to biographical information on foreign military leaders." Joint Publication 2-01 lists 10 different intelligence agencies that provide intelligence support to the warfighters, and some of these organizations provide two or more intelligence support cells [4]. For example, DIA provides Defense HUMINT Services, while the Joint Warfare Analysis Center (JWAC) provides reachback for infrastructure Target System and Critical Node Development.

While it is good to have such separation of tasks of information sources in the intelligence community, it also provides some confusion and difficulty to the planner trying to get a sense of the overall picture and where to go for specific information. The DoD recognized this difficulty and sponsored the creation of the Joint Integrative Planning and Analysis Capability (JIAPC). Although it is still struggling to get off the ground as of the time of this writing, the concept behind the JIAPC is to provide a single source for planners in terms of target characterization (or holistic target characterization). As shown in Figure 10.1, JIAPC brings together elements of the NSA, DIA, and JWAC into a single support element that can provide the holistic characterization of the whole target at once. This concept should help make IO planning easier in the long run and help ensure that cross-sector effects are considered when developing targeting plans.

Similar groupings of intelligence support agencies to the warfighter have already been created, such as the National Intelligence Support Team (NIST), which provides all-source intelligence to the warfighter staff [4]. Having all of the information needed to plan and execute IO missions is an essential

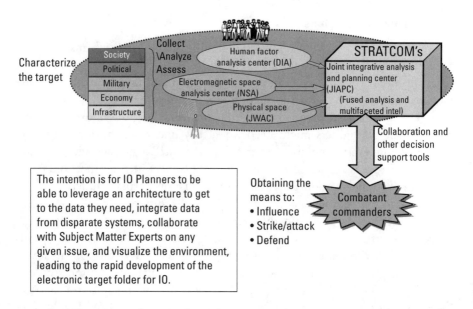

Figure 10.1 Joint Integrative Analysis and Planning Capability (JIAPC) concept overview.

prerequisite to successful missions, and is part of the concept of *information readiness*, described next.

10.1.3 Information Readiness

Information readiness is having all of the necessary information when you need it to accomplish the mission. Classic examples of gaps in required information include having to use tourist maps for the invasions of Guadalcanal and Grenada [5, 6]. As new technology and capabilities continue to enter the force, there is an ever-increasing need for larger quantities of specialized information.

For example, in 1993, the RAND Corporation identified the fact that the Air Force was developing a number of weapon systems with advanced terminal guidance systems that relied on a wide range of target information, but that no one in the Air Force or the intelligence community was tasked to collect that information. As a result of that study, the Air Force inaugurated a policy that required every new system to consider not only its logistics needs in terms of spare parts, but also the logistics of the information it would require in order to operate to specification [7].

Information operations has this problem in spades, as does effects-based operations in general. Attacks against enemy information systems require a wide range of target-related data, as described earlier. Influence operations require a wide range of data on the various groups, messages, and their interactions.

Defense operations require substantial data on the current and projected status of our own resources, as well as patterns of normal behavior.

To address this problem, the U.S. military has undertaken separate but related efforts to provide the information necessary to support IO in particular, and effects-based operations in general. The first effort was the creation of the Standing Joint Force Headquarters (SJFHQs) in each of the combatant commands (CCs)—also known as regional combatant commands (RCCs). Prior to the creation of the SJFHQ, all Joint Task Forces (JTFs) were ad hoc organizations, pulled together from the resources of the CC assigned with a mission, and sometimes supported by the other CCs from outside the region. However, this ad hoc nature of JTFs meant that there was no consistency in how the JTFs were organized, manned, and equipped, which led to planning inconsistencies across various commands and among sequential operations within the same CC. While there were many reasons for the creation of the SJFHQs, one of the benefits was having a permanent staff assigned to prepare in peacetime for future conflicts in an existing go-to-war organization. This permanence provided the assigned resources to begin supporting consistent and coherent information readiness in the region.

The second effort was assigning the development of the operational net assessment (ONA) to support the SJFHQ. Conceived by the U.S. Joint Forces Command, ONA provides the system of systems analysis (SOSA) of the targeted nation that is relevant to the likely missions that the Joint Task Force may be required to perform in the theater of operations. During peacetime and preconflict phases, the CC's SOSA analysts are tasked to develop the ONA to support possible future EBO, including IO. This preconflict information-gathering for possible future use is an essential element of information readiness.[1]

The third effort was to develop the *IO Joint Munitions Effectiveness Manual,* or IO JMEM. The original JMEM was developed to reconcile differences in U.S. Navy and Air Force planning based on different munition effectiveness assumptions. A Joint office was created to run detailed experiments that would definitively specify the effectiveness of every kinetic munition in the arsenal under a range of conditions. The result was a very useful tool for weaponeers supporting traditional military operations. However, no such JMEM equivalent exists for information operations capabilities. U.S. Strategic Command has been tasked to guide the development of the IO JMEM so that current and future planners will have the information they need to effectively plan IO missions with some confidence that the data is sufficiently accurate to provide predictability in applying IO capabilities. The eventual completion of the IO JMEM will be another essential element in information readiness.

---

1. The original form of the operational net assessment is evolving, as the original version was found to be too unwieldy and not as rapidly adaptive to changing situations.

10.2 IO Transformation and Acquisition Planning

This section briefly presents the basic DOTMLPF issues associated with information operations. Every category of DOTMLPF must work together for IO to function correctly in the U.S. military, and therefore planning for the support of IO needs to address each of these issues. The acquisition of IO capabilities includes not only the planning and decisions of the materiel (e.g., equipment) of IO, but also all of the supporting factors that make up the logistics of IO, that is, all of the DOTMLPF factors. However, the analytic and decision support tools to support acquisition planning are currently very rudimentary, and do not have the same capabilities as acquisition for more traditional military capabilities. The last part of this section discusses the need for planning the acquisition of civil cyber-defense.

10.2.1 DOTMLPF

DOTMLPF stands for doctrine, organization, training, materiel, leadership development, personnel, and facilities. The Joint Futures Battle Lab at U.S. Joint Forces Command examines the impact and constraints of all DOTMLPF factors when developing and evaluating transformation concepts. (Each of the Services has a similar set of factors that cover the same topics, if not always using the same names.) Since IO and EBO are relatively new compared to the more strictly kinetic options, the DOTMLPF factors are changing more quickly for IO and EBO than for the other warfighting functions. We will address each of these DOTMLPF factors briefly next.

10.2.1.1 Doctrine

IO doctrine has changed several times in the last few years, including substantial revisions to JP 3-13 and AFDD 2-5, and the creation of DoDD 3600 series. (See Chapter 4 for a more detailed discussion of the recent changes to Service and Joint IO doctrine.) This rate of change will eventually diminish over time, but IO doctrine is likely to continue to change more rapidly than other Joint and Service doctrines due to the evolving nature of IO and EBO. For example, the Air Force definitions and categorizations of IO have become closer to the Joint definitions each iteration of publication, but there are still some significant differences. These differences can usually be traced back to the organizational structure of the Air Force compared to the structure of the Joint community.

Doctrine may be written by one or more individuals, but they are vetted by a community and approved by one or more high-level decision-makers. As a result, what is included in doctrine is almost always a compromise to some degree, and because of organizational inertia, doctrine almost always lags behind the realm of the possible or the efficient. Even so, the rate of change in IO has been rapid and frequent, and progress continues to be made.

10.2.1.2 Organization

The organization of the military to support IO continues to evolve. For example, the IO mission was assigned in 1999 to U.S. Space Command and in 2003 was transferred to U.S. Strategic Command. Moreover, the JIAPC organizational concept described above was developed in 2002 and the first contract to prepare the stand up of the organization was awarded in 2003. Where and how the JIAPC organization will be organized, equipped, manned, and operated is still being debated, although most recently it has been assigned to the Joint Information Operations Warfare Center. Similar organizational rearrangements have been made to try and create an IO range to test IO capabilities and develop the IO JMEM. Such organizational changes will continue to occur relatively more rapidly for IO and EBO than for the more conventional arms for the next several years, but will eventually stabilize as IO and EBO become more commonly accepted members of the military sciences.

10.2.1.3 Training

Training is a big issue for IO. Due to the compartmentalization of many IO techniques, few warfighters have the clearances necessary to know that a given capability exists, let alone have trained with it in peacetime. This lack of peacetime training in IO precludes the ability to train as you fight as well as the opportunity to train on integrating all IO into the rest of the military campaign. Without the ability to train our commanders and staffs with realistic IO capabilities, those staffs and commanders will lack experience with the range of possible IO capabilities and thus be less equipped to apply these capabilities in a synergistic manner with the rest of our national instruments of power. Section 10.3 describes some ways to better incorporate IO elements into training events, testing events, and experiments, while still handling the various classification issues.

10.2.1.4 Materiel

Materiel to support IO is often unique. For example, a bomber delivering kinetic weapons is well understood in terms of flight times, payload, range, time on target, circular error probable, and battle damage assessment. Planning and execution techniques for this and other kinetic options are well understood, as are the patterns by which the materiel is designed, acquired, fielded, and operated. Not so with most IO and other nonkinetic techniques. Acquiring IO materiel could require substantial acquisition of, or interface with, commercial-off-the-shelf (COTS) products. This is because IO must interact with products already in the commercial world, whether part of the civilian infrastructure or in use by foreign militaries. Therefore, materiel acquisition for IO requires more emphasis on researching new capabilities and fielding them to specific organizations for future use, than on acquiring fleets of weapons platforms and munitions. Moreover, due to the rapid changes in the commercial world, the rate at which IO materiel must be updated could be fairly rapid compared to some kinetic materiel acquisition cycles. This would lead to a higher maintenance budget for some potential IO capabilities to stay current.

Another challenge to IO materiel acquisition is the need to thoroughly test the capabilities as part of the acquisition process and to support information readiness (e.g., preparing the IO JMEM). However, some IO capabilities are highly classified because they are fairly fragile. By fragile, we mean that if the enemy were to find out about a certain IO technique, its capabilities could be countered fairly easily. For example, if the enemy knows we are listening in on a certain communication device (like a cell phone), then they will either not use that capability, or use it in a way to distract or disinform. For example, bin Laden supposedly gave his cell phone to his bodyguard to mislead pursuers as part of his getaway plan in Afghanistan [8].

Since all new acquisitions need to undergo developmental and operational testing, new IO materiel also needs to be tested. However, if it is fragile, it cannot or should not be tested in the "open air" where the capability might be detected and therefore nullified. As a result, some testing is performed only in the laboratory or in simulations, which is less satisfactory but may be the only option. The next section discusses some techniques for testing that can help prevent fragile capabilities from being compromised.

10.2.1.5 Leadership Development

Leadership development for IO is a critical need within the U.S. military. Due to the evolving nature of IO and EBO, and the greater scope encompassed by these operations compared to traditional military operations, military education and training has not adapted as quickly as necessary to infuse the concepts of IO and EBO across the current and future military leadership. However, some of the Military Services have recognized this problem and are planning or taking steps to resolve it. For example, the Air Force is planning to educate all Air Force personnel, from airman on up, in EBO, including IO. In addition, the Joint Forces Staff College continues to provide courses in IO for future Joint Staff members. Meanwhile, the Army and Marine Corps continue to emphasize leadership development in influence operations, partly because these are traditional areas of expertise, but also because of the increasing emphasis on the need for such capabilities for the Army and Marine forces stabilizing Iraq.

10.2.1.6 Personnel

Personnel and personnel management are also adapting to the Information Age. The Army has had an IO skill qualifier for a number of years, as has the Air

Force. Personnel management of IO is essential for three reasons. First, a person absorbs the education better when serving on an IO slot just after receiving the IO training. Second, tracking IO skill qualifiers is essential in selecting personnel with the proper experience to head larger units in IO later in his or her career. Third, by the time the person is a general officer, that person should have had at least some exposure to IO in his or her career, or in the careers of those around him or her. The first two have been implemented in most of the Services, but the third step (career exposure to IO before becoming a general officer) has not yet been implemented [9]. Without personnel familiar with the mindset of IO and a thorough understanding of the realm of both the possible and the constraints, the U.S. military will miss the fleeting opportunities to fully exploit the capabilities of IO in today's world.

10.2.1.7 Facilities

Facilities to support IO are also often unique to IO and essential to the success of IO. As mentioned in the preceding sections, much IO materiel is unique, as are IO organizations. As a result, unique facilities are usually required to house, secure, and operate IO capabilities. For example, the JIAPC described earlier will have extensive communications bandwidth in order to facilitate large volumes of data to support extensive responses to requests for analyses. Other types of facilities include the IO ranges, which will be essential to the development of the IO JMEM and to the testing of more fragile IO capabilities.

Overall, IO has a number of DOTMLPF issues that make IO and EBO unique when compared to more traditional military capabilities. All of these DOTMLPF factors must be considered when planning any military transformation, and if any part is missing, then it is likely to be a source of problems for the commanders and operators in the field.

10.2.2 Planning Methodologies for IO Acquisition

Due to the time delay from fiscal planning to actual expenditure, most fiscal decisions tend to have multiyear planning horizons. This leads to two features common to fiscal decisions. First, the fiscal decisions being supported tend to be broad category multiyear investment decisions. Second, due to the uncertainty of looking that far into the future, the level of detail of the information required or available to support the decision is not very deep.

For example, most people would consider it ridiculous if one proposed to apply a very detailed linear program to allocate resources by dollar and day to fiscal expenditures 5 years from now. So much can change in technology, world events, domestic politics, and the economy that to attempt such detailed planning for upcoming years would be a fruitless exercise. Any assumptions made now for 5 years in the future are likely to be false.

Therefore, budget allocations for 5 years into the future tend to be made in increments of tens to hundreds of millions of dollars for investing in relatively broad areas of desired capabilities. Some long-lead items, like aircraft carriers, tend to require multiyear expenditure plans. Even the fiscal planning cycle tends to list major programs in priority in millions to tens of millions of dollars.

Long-range fiscal planning and decision support for IO follow this same pattern. Although many IO capabilities need to consider operating in the millisecond time frame, fiscal decisions for even these programs are made in the multiyear range, in increments of many millions of dollars.

Moreover, IO fiscal decision-making involves a few additional considerations, mostly related to the classification of most IO programs. Due to the highly compartmentalized nature of the IO business, it is inherent that some overlap in capabilities will occur. Fortunately, each of the programs tends to add something unique to the set of overall capabilities, but the fiscal decision-making is not detailed enough to be able to identify and rectify any actual overlap.

For fiscal planning support, the level of detail required in the tools and analysis is directly proportional to the level of detail required in the decision. If the decision is for broad-brush allocation of resources across various capability areas, then tools and techniques that apply to broad tradeoff categories will be appropriate. Highly detailed models of intricate computer techniques or influence operations will not be appropriate for that level. Conversely, if one is attempting to determine how much it will cost to design, develop, implement, and support a particular IO capability in any of the three realms, then more detailed models of the specifics of that type of capability will be required to help support and justify the acquisition decision.

For fiscal decisions and deliberate planning, more time is usually available for analysis, which in turn allows for more detail and cases to be considered. Each case to consider includes a number of assumptions and input factors that need to be prepared, whereupon the analysis is performed and the outputs are analyzed. The more detailed the models and algorithms used for the analysis, the more time is required to prepare, execute, and analyze the results of a run. In addition, the level of detail in each remaining case tends to increase over time, as the number of cases considered in detail decreases. Therefore, early in the deliberate planning and budget analysis processes, the scope is large and the number of cases is large, but as the deadline for completion of the analysis approaches, a smaller number of cases tend to be examined in more detail. Even after the analysis is completed, additional excursion cases tend to be run as relatively minor variations on the ones already performed.

Overall, the first rule of thumb for supporting IO fiscal planning is to keep the decision support data at an appropriate (not very detailed) level, with time measured in months or years, and dollars grouped in millions or tens or hundreds of millions. The second rule of thumb is to be aware that the high classification level will make it difficult, if not impossible, to be efficient.

As of the time of this writing, there do not appear to be any methodologies or tools unique to IO being used to support fiscal decisions for the acquisition of IO capabilities. Due to the comparative newness of IO to traditional military operations, the military science of IO is still evolving, and the number and capability of models and simulations applicable to the broad field of IO are still very limited. There are as yet no models of IO comparable to TacWar, Thunder, ITEM, Combat Evaluation Model (CEM), or other combat models used to define requirements for conventional munitions and conventional system acquisition. Nor are there training models similar to the Corps Battle Simulation (CBS), Air Warfare Simulation (AWSIM), and the rest of the Joint Training Confederation. Moreover, claims that such traditional models adequately represent such considerations have been discounted for lack of credibility. One reason for the funding of the Joint Simulation (JSIMS) and the Joint Warfighting Simulation (JWARS) programs was to help account for representing war in the Information Age as opposed to the Industrial Age. However, the JSIMS has been cancelled and no new contract has yet been let, while the JWARS contract was also cancelled, but has been given at least a partial lease on life by USJFCOM J9.[2] However, progress is being made in the various stages of IO planning (deliberate, crisis action, execution, persistent) and mission execution. As these planning tools evolve, their approaches are likely to be applicable (with modification) to the fiscal and acquisition applications.

As a result, fiscal decisions for IO and IO-related programs will continue to use the tried-and-true techniques of decision analysis, spreadsheets, briefings, and reports. These simpler fiscal decision support techniques can usually be applied successfully to address the budgetary and programmatic issues associated with IO programs. There are just no good models yet to address the tradeoffs among alternatives in the land, sea, air, space, and information domains. Such a tool would be useful to support fiscal decisions and the tradeoffs among the various capabilities proposed for funding.

Once such models do exist, however, their representation of the strengths, weaknesses, and synergies among the various domains needs to be validated. For example, deep fires can be conducted and/or augmented by land, sea, air, and Special Forces assets. If the analytic tool does not represent all these capabilities, then the assets of one domain may be credited with unique capabilities that are not, in reality, unique. This need for adequate representation of strengths, weaknesses, and synergies is essential for tradeoff decisions and will have an even

---

2. JWARS does, however, have the basics required to be able to represent many IO features in the future because it bases its simulated decisions on perceptions rather than model truth, and tracks the passing of information in its communications model.

greater impact on decisions related to information operations assets. This is because analysts and decision-makers are less familiar with the actual capabilities and limitations of IO assets and will be more likely to rely on the model results when their experience base is lacking.

10.2.3 Acquiring Civil Cyber-Defense

As described in Chapter 2, our Information Age society is more dependent upon, and interconnected with, our civilian information systems than ever before. This leads to significant vulnerabilities in our civilian society, commercial activities, and open governance. Civil cyber-defense is clearly a defense-related issue, but it is unclear who should pay for increasing the security of our civil information infrastructure, who should be responsible to make sure it improves, and how such a program should be run.

One hypothesis is that market forces alone will provide sufficient protection to the civilian information infrastructure. The assumption is that as the demand for security hardware, software, and techniques increases, companies will arise to provide these goods and services, and the problem will be solved. Unfortunately, this solution has not yet appeared, as companies seek to cut costs, cyber-security is one area where allocating funds has not been sufficient to date. Usually, a company is either consciously or unconsciously accepting risk until a disaster happens, whereupon more resources are allocated to making sure the problem does not happen again.

Cyber-security is complicated, and as stated previously, the attackers currently have the advantage. To have good cyber-defense, a company needs a good cyber-security professional, but few companies are willing to pay for such skills. They rely on "good enough" in-house or outsourced computer support with some security, only to find that such support really was not good enough. Since most senior managers are not versed in cyber-technology, they find it difficult to determine who is sufficiently competent and who is not. As a result, most companies believe they are more secure than they actually are. (Note that the federal government is often even more at risk than the commercial sector since they find it even more difficult to hire and retain competent cyber-security personnel [10, 11].)

Efforts are underway to either incentivize organizations to improve their cyber-security, or to disincentivize (punish) organizations that have weak cyber-security. For example, organizations with lax cyber security that allows their systems to be used to attack other systems could be threatened with lawsuits based on "cyber negligence" [12].

In the long run, we are likely to see cyberspace divided into enclaves, where more secure enclaves require everyone to be identified and there is no anonymity in which to hide nefarious acts [13]. This would be in contrast to the free fire zone of the current Internet, also known as the Wild Wild Web. Individuals who are tired of the problems that come with unrestricted freedom will opt for maintaining their cyber-residence in a secure enclave, and let the unrestricted zone degrade to whatever level it will reach due to global neglect. This would, in the long term, solve the cyber civil defense problem, but also tend to preclude some of the great value of the Internet to bring together diverse communities. Moreover, if each nation can build its own secure enclave that all its citizens must use, then the freedom and cross-fertilization provided by the current Internet will disappear.

10.3 IO in Training, Experimentation, and Testing

This section will describe the unique challenges IO presents when applied to training, experimentation, and testing events. The logistics tail of IO includes the training of personnel and organizations in IO, the testing of IO strengths and vulnerabilities in the acquisition of IO and non-IO assets, and the inclusion of IO in experimentation to support the DoD's transformation process.

Each of these three types of events has a different purpose and set of constraints. Training involves a training audience participating in an event to exercise their previously learned skills and to experience new situations to better prepare for combat or crisis. Experimentation involves the creation, analysis, and testing of hypothesized concepts for improvements in how to operate, including new items, uses, or combinations of doctrine, organization, training, materiel, leadership, personnel, and facilities (DOTMLPF). Testing involves subjecting a system to a set of conditions to determine whether it performs to specification. The system under test may be an IO asset, or an asset that might be subjected to enemy IO, or an asset that might be used synergistically with other friendly IO assets. Each type of events has its own issues to address and may have different solutions to implement.

10.3.1 IO in Training

The U.S. military training community uses the maxim "train as you fight." This means that for training to be good preparation for actual combat or conflict, the skills and procedures learned during training should match as closely as possible the conditions one would face in real-world events. If training does not provide that type of learning experience, the training audience may come away with the wrong lessons learned.

If units do not train with IO capabilities and procedures in peacetime, they will not know how to properly plan for and use these IO capabilities in wartime. Without some prior experience in training events with IO capabilities and vulnerabilities, real-world operations will be hampered by lack of knowledge of, or experience in, IO capabilities and vulnerabilities.

Offensive IO is rarely included in training events, however, for two reasons. The first reason is that such IO capabilities are often highly classified. Since most training events are classified only to the secret level (even if a more highly classified intelligence support facility is included in the event), IO is not played in most training exercises. Further, few personnel in a theater of operations traditionally have the clearances necessary to know what IO capabilities are available, let alone how they could be used to support the overall theater plan [14]. As a result, the benefits and constraints of actual IO planning are often not rehearsed in peacetime, which can lead to problems in wartime or crisis.

A second reason why offensive IO is not commonly included in training events is the fear that its successful application will preclude the achievement of the training objectives. For example, if the purpose of an exercise is to determine how the training audience (Blue) will react to a major problem or setback, this typically requires a strong, capable Red or opposing force (OPFOR). If, however, a successful Blue IO action has caused the OPFOR to become paralyzed due to lack of information or lack of confidence in their information, the actual training objective of the exercise may not be met. Similarly, if the Red IO play causes the Blue side to become incapacitated, this could also preclude the achievement of the purpose of the exercise. Therefore, IO capabilities for either the training audience or OPFOR cannot be allowed to dominate the event and preclude the achievement of the training objectives.

Even so, the potential effects of IO in an exercise can be mitigated by various techniques. For example, influence operations can usually be played in training events through the use of *role players*, who can react in a way to keep the action going, but demonstrate that the influence operation had some effect on the outcome. For example, if a mob is forming, a PSYOP action may reduce the size of the mob, but still provide sufficient mob activity to stress the Blue force. However, the problem with using U.S. role players in training events is the tendency to mirror-image the enemy's thought processes as a reflection of our own. As we have found in the Middle East, this is often not the case. To avoid this problem, training events for units about to be deployed to Iraq often include Iraqi expatriates as role players to provide a more realistic representation of the benefits and pitfalls of influence operations at the tactical level [15]. Unfortunately, it is more difficult to provide a similar realistic representation of the benefits and pitfalls of influence operations at the operational and strategic level in higher-echelon training events because of the lack of expatriate role players who have higher-level command experience. This might be a good use of Coalition partners, however, who might be able to provide opposing force personnel with appropriate experience in higher-level exercises.

Representing IO attack and defense faces similar problems. For example, if one side's OPSEC plan fails, the other side can gain a significant advantage. If that degree of knowledge gained by the opposing side will completely unbalance the event, the control cell must take steps to mitigate the magnitude of that OPSEC leak. It might delay the time of discovery to reduce its impact, set up an event that will enable the team to learn that its plans have been discovered, or set up conditions that will prevent the discovering side from taking full advantage of the information. In a similar manner, the success or failure of a military deception operation can create a significant impact on the course of the training exercise. The impact of a successful military deception operation can be devastating for the side that was fooled. It is fairly difficult to mitigate the impact of such a significant event, and the control cell may again need to create events to help keep the training event challenging. Note that the purpose of the control cell is to keep the play sufficiently balanced so that the exercise provides the opportunity for all of the training objectives to be met. The purpose is not to run a training event like a game to see which side wins [16].

The representation of computer network attack and computer network defense in a training event raise additional issues. For this reason, IO should not be played on the exercise control network. The exercise control network is what creates ground truth for the event, and neither side's IO capabilities should be allowed to interfere with the representation of ground truth. Otherwise, the lessons learned will be corrupted. Section 10.3.2 describes how computer network operations (CNO) can be successfully represented in training and experimentation events.

10.3.2 How to Include CNO in Exercises and Experiments

One of the problems frequently raised in the design of exercises and experiments is how to let CNO operators have "freeplay" and still ensure that the exercise objectives are met or the experiment's hypotheses are adequately tested. The primary concern is that if the use of CNO by one side (Blue or Red) is too successful, it will dominate the results of the exercise or experiment and preclude achievement of the event's objectives. A secondary concern is that if CNO is played on the network that provides ground truth for the event, this will cause the information about ground truth to be corrupted, precluding not only the achievement of the event's objectives but also wreaking havoc on the event's network. This section focuses on how to represent CNO in exercises and training events without fear of shutting down those events.

Throughout this section, we use the term *event* to refer to a training exercise or concept experiment. The training audience is the Blue Cell in an exercise, or the Blue Force in an experiment. The Red Cell is the opposing force (OPFOR), and the White Cell is the control cell. The White Cell sees the ground truth during the event, while the Red and Blue Cells see only the information that should be available to each side.

We recommend the following four steps for including CNO in events:

1. Ensure that the cells to be targeted by the CNO methods of the opposing side are physically separated and forced to use electronic communications.

2. Apply all CNO techniques against an actual or simulated network that is *separate from but parallel* to the computer and communications network used to run the event.

3. Place the White Cell in charge of implementing selected outcomes of CNO actions received from this parallel network to be entered into the event's play.

4. After the event, interview participants to determine what would you have done given certain CNO actions, and collect lessons learned.

The first step is to ensure that the player cells are physically separated and unable to communicate by any nonelectronic, unmonitored means (e.g., voice or written messages). Since the mechanism of most CNO methods is to affect information content or flow to influence the perceptions of the decision-maker, it is important not to enable players to easily resolve information discrepancies through direct contact. If information flow and content are channeled through monitorable communications channels, CNO can be applied against the player cells with some likelihood of influencing decisions and actions in the event. Note that this means not only that the Blue Team must be separated from the Red Team, but that Blue Cells that may receive an IO attack must also be isolated from one another (as must Red Cells) to ensure that differences in perception can occur within the same team.

The second step is to ensure that all CNO techniques involved in the event, whether real-world or simulated, are applied to a network that is *parallel* to the network supporting the event itself. One does not want the CNO actions to disrupt the ground truth that the White Cell is using to manage the event. After-action reviews (for exercises) and hypothesis findings (for experiments) are also very difficult to achieve without a protected ground truth from which data can be drawn after the event. To accomplish this, a parallel information network must be developed, whether real or simulated, and all CNO actions are taken within or against this network. When a particular type of attack succeeds, the report of that success is sent to the White Cell, which determines whether that effect (or how much of that effect) should be included in the play of the event and when. By affecting only the parallel network and measuring success or failure against that network, the CNO effects do not affect the play of the event itself unless approved by the White Cell.

Thus, the third step is for the White Cell to determine whether to allow the effects achieved on the parallel network to be played, when, for how long, and to what degree. For example, if Blue has achieved an in-line tap on a simulated Red network, the White Cell can allow some of the raw traffic of communications to be made available to the Blue Team, or have one of its analysts provide only a subset of the relevant data to the Blue Team. Since part of the White Cell's job is to meter the pace of events (and make the event challenging to the Blue side), it can use these CNO results to increase or decrease good or bad events for each side. An exercise should not be so overwhelming that Blue does not achieve its training objectives, nor so easy that team members achieve their training objectives without effort. In experiments, the White Cell needs to ensure that the play is balanced so that all of the hypotheses can be tested before the end of the experiment. By providing the achievements of the CNO attacker or defender on the simulated, parallel network to the White Cell, the effects can be metered into the event by the White Cell so that CNO successes do not dominate the event.

The fourth step takes place at the end of the event, when all the data necessary to support the AAR or experiment findings are collected to ensure that the objectives of the event were met. At this point, it is time to evaluate what could have happened in the event, and how each side would have reacted had certain events occurred within the play.

When CNO is involved in the event, all the successful CNO actions should be laid out on a timeline that also includes the event's major milestones. Successful CNO actions that were allowed by the White Cell should be highlighted on the timeline, including those whose effects were delayed. The Blue and Red Cells are either jointly or separately presented with this information to show the realm of what could have happened had all of the successful CNO actions been allowed by the White Cell. The personnel in the affected cells are also interviewed to determine what they would have done had these successful actions been allowed by the White Cell. At this point, the observers and analysts now have sufficient information to show how CNO play could have affected the play of the event, when it actually *did* affect the play, and how each of the cells believe they would have responded to such stimuli. Thus the effects of CNO on each side can be obtained, without losing the opportunity to ensure the achievement of training objectives or hypothesis testing [17].

10.3.3 IO in Experimentation

There are many types of experiments that one can undertake, including analysis, seminars, workshops, wargames, limited objective experiments and large-scale experiments [18]. In the first three types, the representation of IO, like most other aspects, is relatively limited in detail. This is because the time frame for most experiments is at least 5 years into the future, more often 10, and sometimes 15 to 25 years away. As a result, the level of detail of every functional area is very limited, including that of IO. Therefore, the representation of IO does not have to be very detailed in analysis, seminars, and wargames.

In limited objective experiments (LOEs), the purpose is to put many of the major pieces together to see how they interact. In the Information Age, IO is likely to be a significant piece in most future concepts. Therefore, the play of IO may have a significant effect on the findings of the LOE, and therefore the realistic play of IO in experiments is more important in LOEs than in the first three types of experiments.

In a similar manner, large-scale or major experiments should also include a realistic play of IO. IO is likely to be a major player in the experiment's hypotheses, if not as a major hypothesis, then as a possible facilitator or block to the success of a given concept being considered. Due to the high degree of interactions among elements in major experiments, such experiments tend to be manpower-intensive and entail more detail than the other four types. Both LOEs and major experiments tend to take on some of the aspects of a training exercise, at least in terms of the opportunity for IO to dominate the event and possibly preclude the achievement of the experiment's objectives. Section 10.3.2 described how to avoid this.

10.3.4 IO in Testing

Testing is part of the acquisition process and includes both developmental testing and operational testing. In developmental testing, the developer (such as the contractor and/or the government's program office) tests the system to ensure that its performance conforms to requirements or specifications. In operational testing, the types of people who will operate the system in the field operate the system under test to ensure that its performance is both effective and suitable [19].

The U.S. military's reliance on networked C4ISR systems continues to grow. With this increased reliance comes increased potential network vulnerabilities. To ensure the information security of U.S. systems that will operate on a network when fielded, developmental testing and operational testing should include threat representations against such systems. Moreover, the combination of systems provides additional vulnerabilities not inherent to individual systems. These system-of-systems vulnerabilities must be tested as well as the vulnerabilities of each component system.

Many of the U.S. military's test and evaluation master plans (TEMPs) do not include IO tests for two reasons. First, the TEMPs were written 10 to 15 years ago, and the IO threat was not well understood at that time. Second, most systems were not originally designed for IO defense in depth. The assumption was that if the perimeter of the communications system was secure, then the systems within did not require their own IO defensive capabilities. Understanding of the IO threat has dramatically increased, and the testing community is just beginning to realizing that IO should be part of every system's developmental and operational test event.

Support of IO in testing new systems faces three major challenges. The first is that testing new systems is always politically sensitive. Second, most classified IO techniques cannot be used in the open air without the potential for compromise. Third, tests that include IO include the familiar problem of potentially dominating the test event, thereby precluding the testing of all relevant features.

For the first challenge, testing new equipment in a realistic IO environment may be considered unfair to the system under test. Most new systems undergoing developmental testing tend to be fairly fragile to begin with, and hitting them with IO is likely to cause the system to fail the test. As a result, most program offices will not allow realistic IO conditions to be present in a test to protect their system from unfair evaluation. A commonly used argument is that unless every system has been tested under the same IO conditions, then no one system should be singled out for such a test. The solution to this problem is to incrementally include IO in developmental and operational tests over time, so that IO becomes an accepted part of the testing process for all systems. Progress has been made in the last few years, and such tests are now actually being performed on some systems.

The second challenge is that the IO capabilities being tested or being used to support testing other systems may be highly classified. Due to the highly classified nature of most IO capabilities, testing usually cannot be done in the open air or in the wild without extensive precautions being taken to preclude compromise. Most testing of IO systems needs to be done in isolated laboratory environments where the risk of compromise is zero or nearly so. However, without realistic test environments for IO systems, the confidence that such systems will work as intended in the real world is not high. As a result, IO systems need to be subjected to similar types of testing as is currently done for kinetic systems. Such an IO test range is currently being developed, but is not yet operational at the time of this writing [20].

The third challenge is similar to the dominating aspect of successful IO in training and experimentation. Testing events are expensive, and the tester cannot risk not meeting the test objectives due to the failure of one aspect. As a result, IO should be included (at least in the near term) as a passive participant in the testing event until the other test objectives have been met. During the last few hours of the test, after the other test objectives have been achieved, the IO capabilities should turn active, exploiting penetrations achieved during the passive period, and otherwise wreaking havoc on the system under test.

Information security threat representations in operational tests must be realistic, yet must not disrupt the achievement of the rest of the test and evaluation objectives. Any network intrusion threats should be represented in such a manner that these threats have no apparent effect unless specifically called for in the test plan. There are six steps to applying CNA in the testing environment:

1. Test period setup;
2. Network access;
3. CNA software intrusion;
4. Activation of active CNA effects;
5. Postevent analysis and presentations;
6. Archiving results.

10.3.4.1 Step 1: Test Period Setup

Prior to the test, the intrusion hardware and software suite should be set up in a control location of appropriate classification. Since the CNA control suite can be located anywhere, it can be housed at a location convenient to all parties.

10.3.4.2 Step 2: Network Access

Access to the target network is essential. The CNA software will need at least one access point into the target network and continuous contact back to the CNO control station. The key question is how to create a plausible network access point early in the test that will be acceptable to the Blue forces during the after-action review. Analysis of the best access approaches must be done prior to the test event to ensure that access will be available to the CNA software. At worst case, the test can employ a final fallback position of a scripted access to a network target point to ensure that the CNA software is demonstrated. If a credible network access method is not successfully and clandestinely applied during the test, the Blue security personnel can take credit for having blocked the attempted access. The scripted access play from that point forward is to determine what would have happened if the access had not been detected and blocked.

10.3.4.3 Step 3: CNA Software Intrusion

During the test scenario, the Blue forces will set up and operate their networks, including the system under test. CNA intrusion will occur as early as possible, and continue throughout the course of the test. The CNA software will perform the following three actions:

- Map the target network and report the map to the CNA software control center. Any modifications of the target network by Blue personnel over time will also be reported to the CNA software control center.

- All intrusion actions, and the information gathered by those intrusions, will be time-stamped and logged for later review, analysis, and presentation after the event.

- Toward the end of the test, the intrusion software will leave electronic calling cards in selected locations throughout the target network. These calling cards will not modify or disrupt any data or network parameters that could disrupt the test. If detected by Blue personnel, observers will note the fact and record reactions to the discovery. If Blue reactions threaten to disrupt the test, observers will report this to the CNA software control center, which will then terminate that portion of the CNA test.

10.3.4.4 Step 4: Activation of Active CNA Effects

Depending on the test objectives, some active CNA effects may be included in the test. These active effects should only be included after the achievement of all of the other test objectives. Once the other test objectives have been achieved, the previously inserted intrusion software can be activated in the target network to determine the effect it would have on the system under test, and how well the system under test can handle the assault. This step should not include any destructive means. For example, if an attack involves manipulating data, a copy of the data needs to be copied offline before the attack is launched. This test provides a good exercise for the network defenders and provides data on how well the system under test could respond to such threats.

10.3.4.5 Step 5: Postevent Analysis and Presentations

Upon conclusion of the operational test, the recorded CNA intrusion data can be analyzed and prepared for presentation to various audiences. Some presentations can be held at the collateral level, while others may be at the highest level necessary to explain what actually happened and why. The presentation of results should include (as appropriate to the audience) how the intrusion was attained, the degree of success, and the information gathered (or other effects achieved) by the intrusions over time. The time of selected key events in the test should be used as reference markers for Blue personnel to comprehend the level of penetration relative to key operational test events. Recommendations on indications and warnings of similar threats, and procedures for how to help prevent such attacks in the future, will also be provided at appropriate levels of classification.

10.3.4.6 Step 6: Archiving Results

Summary data from the event should be retained at an appropriate level of classification if deemed valuable by the customer. Such archived information can be used as briefings at the appropriate level to further customer objectives.

For a test event that desires to use IO attack or defense, several unclassified tools have been successfully used to support testing events and general Red teaming of potentially targeted friendly networks. Four such tools are Maverick, the Information Assurance Test Tool (IATT), the Network Exploitation Test Tool (NETT), and the Internet Attack Simulator (IAS).[3] These network attack simulators include a set of primarily open-source hacker techniques combined in a GUI designed framework for nonhackers to use. The advantage of these systems is that attack techniques seen in the wild can be included as scripted elements, which are then agreed to both in content and in timing of application during the event. All events are time-stamped, and a log of the results is stored to support after action review.

In summary, IO needs to be included in training, experimentation and testing, but care must be taken to include the effects of IO, especially CNO, in the right way. Otherwise, the effects of IO can potentially dominate the event, and thereby preclude the achievement of the event's objectives.

10.4 Legal Issues of IO

This section provides a brief discussion of the legal issues of IO. Due to the newness of the field, and due to the difficulty in proving the source of an act, there are no clear definitions of acts of war, acts of force, or hostile acts for IO in the international arena. Conversely, there are a number of misperceptions of assumptions about the use of IO in peace and conflict, which we discuss in Section 10.4.2. Section 10.4.3 focuses on how to help planners and commanders understand what is legally allowed during both planning and execution monitoring. Section 10.4.4 briefly discusses the inherent conflict between Title 10 (military) versus Title 50 (intelligence) organizations, which have been exacerbated in the Information Age.

10.4.1 What Is a Hostile Act?

In more traditional military operations, a hostile act has become fairly well defined. For example, locking targeting radar on a ship or airplane constitutes a hostile act, as it is a precursor to an attack and not necessary in peacetime [21].

---

3. The first three were developed by General Dynamics, while the fourth derived from a Web-based version of Maverick developed by Atlantic Consulting Services.

Of course, firing on enemy forces or other national assets is considered a hostile act. Depending on the circumstances and magnitude, such hostile acts can be considered an act of war.

However, it is more difficult to unambiguously define an act of force or war in the realm of IO, and often more difficult to even prove the perpetrator of a detected hostile act. For example, breaking into the computer system of a nation and stealing information is clearly a hostile act. The Chinese military has been identified as the likely source of a number of recent break-ins in U.S. defense industry company computer networks [22]. This is clearly a hostile act, but it is difficult to prove without a reasonable doubt that the Chinese military is behind these attacks. So the response has been to focus on hardening our civilian cyber-defenses and ignoring clearly hostile acts until a clear perpetrator can be unambiguously identified.

If a nation undergoes a serious denial-of-service attack so that the nation's economy is severely disrupted, this might be considered an act of war. China and Russia are very concerned about such a threat to their respective economies. One Russian academic published the position that "Russia reserves the right to respond to an information warfare attack with nuclear weapons" [23].

A different set of actions is considered in peacetime versus in wartime. Just as one nation dropping bombs on another nation is consider an act of war, one nation inserting logic bombs on another nation's cyber-infrastructure could be considered an act of war if the effects were widespread (and assuming the source could be proven or highly probable).

Since most IO capabilities are nonlethal in nature, they are more likely to be selected as the option of choice in peacetime. Moreover, in these times of increased concern for collateral damage and minimizing casualties, IO capabilities may become more frequently selected by the United States and other Western nations during future conflicts. At this time, there is a greater need to make the expected outcomes of IO capabilities more predictable, repeatable, and validated so that commanders and planners can best understand the risks involved in applying these relatively new capabilities. In addition, there is a need to ensure that commanders and planners understand what is and is not legally allowed, which is the subject of Sections 10.4.2 and 10.4.3.

10.4.2 What Is Legally Allowed in IO?

U.S. planners are very concerned about planning and executing only those actions that are legal. Almost every military member wants to be a good representative of the United States toward the rest of the world. Therefore, it is important for planners to understand what is legal and to not avoid actions due to a misunderstanding, such inaction could result in severe operational consequences.

For example, let us take the case of two hypothetical legal questions:

- Is it legal to assassinate the leader of an enemy nation in wartime?
- Is it legal to gather intelligence on non-U.S. citizens who are not permanent residents in the United States?

The correct answer to both questions is yes, but there were two recent cases in which military personnel believed the answer was no, which led to inaction that resulted in severe consequences. We refer to these two examples as hypothetical cases because the actual cause for *why* these two actions were not taken has been extensively disputed. However, these two hypothetical cases provide good examples of the potentially severe consequences of assuming that a legal action is illegal.

In the first case, it was reported that during the liberation of Afghanistan, there was an opportunity to strike the Taliban leader Mullah Mohammad Omar with a UAV-launched missile, but that someone on the staff stated that the action would be illegal [24]. Therefore, Omar was not attacked and he is still at large many years later. This legal assessment was reportedly based on the fact that there is a law that forbids the assassination of leaders of foreign nations [25]. However, this prohibition does not apply in wartime. The executive order against assassinations of opposing leaders was a directive given to the U.S. intelligence community, was only intended for nonwartime situations, and was originally designed to preclude rogue operations against foreign leaders opposed to the United States [26, 27]. As a result, the attack on Omar would have been legal. By the time the liberation of Iraq was launched, this was well understood by planners since there were several attempts to bomb Saddam Hussein [27].

In the second example, the military intelligence group called Able Danger reportedly had identified Mohammad Atta and three others as a potential threat prior to 9/11. The team wanted to pass the information they had gathered to the FBI for follow-up. However, it was reported that a legal assessment was made that since Atta was already in the United States legally, sharing intelligence about him was illegal. Again, this was an incorrect assumption [28]: "The prohibition against sharing intelligence on 'U.S. persons' should not have applied since they were in the country on visas and did not have permanent resident status."

As a reminder, there is significant question as to whether either example occurred as described. That is why they are presented here as being hypothetical. The point is that when someone during planning or execution incorrectly assumes a legal action is illegal, severe consequences can result. Erring on the side of inaction due to misunderstanding the legal situation can be just as disastrous as erring on the side of action.

However, the debate about what is legal in IO is not limited to misunderstandings of the law. There are also differences between what the United States considers legal in IO and what other nations, including the United Nations, consider illegal. For example, U.S. government lawyers prepared a 50-page legal discussion paper (or white paper) outlining international law as it applies to information operations [23]. Although the final conclusion of this document states "There are no 'show-stoppers' in international law for information operations as now contemplated in the Department of Defense," this assessment is substantially different from the U.N. view on the subject.

The U.N. argument is described in the *United Nations Manual on the Prevention and Control of Computer-Related Crime* as follows [29]: "An attack through a network that crosses neutral territory, or using a neutral country's satellites, computers, or networks, would infringe upon that neutral's territory, just as would an overflight by a squadron of bombers or an incursion by armed troops. The attack would thus be considered illegal and, perhaps, an act of war against the neutral." The rationale is that electrons passing through cyberspace constitute a violation of national sovereignty if they pass through a neutral nation to reach the targeted nation. Thus, if military commanders and planners feel compelled to follow the guidance in this *U.N. Manual*, it could greatly hamper future military operations.

In its argument, the U.N. is attempting to use precedent or analogy to equate the movement of electronic signals in cyberspace to the physical movement of troops across the border. In the opinion of this author, however, a more appropriate analogy or precedent for a legal position would be based on comparing the electronic signals passing through cyberspace to electromagnetic signals in the ether.

The passing-electron analogy is flawed because the electron from the United States is not directly passing from the United States through the neutral nation to the targeted nation. When an electronic signal is transmitted from one location to another, the same electron is not moving that whole distance. In many cases, the electron stops after only a short distance, and other electrons continue to carry the signal. It is the flow of the electromagnetic signal through a conductive medium that is moving and not a specific electron. Therefore, a more appropriate analogy of cyberspace to the physical world is the transmission of electromagnetic (EM) signals across national boundaries, or the signals carried on a telephone line. In most cases, the passage of such signals are not controlled when crossing neutral national boundaries (subject to frequency interference issues), and therefore are not subject to laws associated with the physical violation of a neutral's sovereignty. Thus, if an EM signal is sent from one nation across a second nation to a third nation, the nation in the middle is not held liable for the signal passing through its territory. In a similar manner, signals sent in cyberspace should not be considered any more a violation of national sovereignty than the passing of EM signals, or phone calls that are routed through a third nation. If someone makes a disturbing phone call from France to the U.S. that happens to pass physically through England, that does not make England responsible for letting the phone call go through.

Communications technologies and networks are governed by a set of international agreements where third-party interference in the transmission of information is precluded, and conversely, neutral parties involved in the transmission avoid the responsibility for content. International agreements exist for electromagnetic communications (in the air), electronic communications (on a wire), and physical communications (such as international mail). Therefore, it would not be appropriate for France to intercept and open mail traveling by train or aircraft from England to Italy across France. Conversely, if a letter bomb were sent from England to Italy, France would not be held accountable, nor would its sovereignty be considered to have been violated. Even if the letter were not a bomb but a political statement, the passage of that message across French territory would not be considered endorsement of that political message simply because it passed through France. The same rules, in this author's opinion, should apply to electronic (Internet) communications.

In general, U.S. planners must realize that what is or is not legal in IO is likely to change over time, and that there are also likely to be disagreements between what the U.S. perceives as legal and what other nations may perceive as legal in IO. This is another reason why having a complete audit trail of planning considerations, including legal considerations, is essential to modern planning tools.

10.4.3 How to Help Planners Know Up-Front What Is Allowed

One way to help handle this problem is to define standardized rules of engagement (ROEs) for IO, as described in Section 5.2. This approach includes definitions for measures of effectiveness (MOEs), ROEs, and constraints and restraints that can be quantified and tracked from the higher command authority (HCA) all the way down to the individual planner. This helps in a variety of ways, including allowing the political authority to visually comprehend the effects their ROEs will have on the space of military options.

In addition, having the legal issues considered become part of the standard audit trail for the plan will help ensure that the rationale is fully documented and ready for review. If a flawed legal assessment has been captured in the planning artifacts, then the review process of why a certain desired action was not allowed will be highlighted for review. At the moment, the planning tools described in Chapter 9 allow for text string comments to be entered, but the discussion of legal issues beyond the ROE bins described earlier currently need to be entered as comments in the elements of the plan or CAEN. It would be useful for the planning and military legal community to have a section identified as legal highlights in addition to the ROE bins for the purpose of highlighting to reviewers the legal issues that were considered, their rationale, and conclusions.

10.4.4 Title 10 and Title 50 Legal Issues

The final legal issue discussed in this chapter involves what are called Title 10 versus Title 50 organizations. The U.S. Military Services are governed by Title 10 of the U.S. Code, which prescribes the organizations and other requirements responsible for manning, equipping, and training the land, sea, air, and any other military capabilities. The U.S. national intelligence agencies are governed by Title 50 of the same code, which prescribes the organizations and other requirements responsible for intelligence operations. During Industrial Age warfare, this distinction was usually sufficient to keep the responsibilities (and funding) of each area separate. However, due to the significant expansion of information sources and capabilities in the Information Age, and in the anticipated expansion of capabilities in information operations, the distinction between intelligence operations and military operations has blurred substantially.

For example, let us assume that a Military Service has an offensive IO capability. Joint military doctrine states that the Service needs to perform intelligence preparation of the battlespace (IPB) to understand the target and environment of a potential conflict. However, the act of gathering that information is claimed by a Title 50 national intelligence organization, which may refuse to let the military perform its necessary IPB. If this hypothetical situation were to play out, the United States could be attempting to perform a military operation without sufficient intelligence.

Conversely, let us assume that a national intelligence organization is gathering information about a potential adversary. Due to the nature of modern information systems, more complete information can be gathered during combat operations if the intelligence agency were allowed to perform some activities currently assigned to the U.S. military. During wartime, however, such interactive intelligence-gathering capabilities may actually fall under the mandate of Title 10 organizations.

Thus, the distinction between the responsibilities of Title 10 and Title 50 organizations is no longer clear, leading to boundary disputes and funding concerns. As a result, cooperation between the two communities has sometimes been poor in the past, leading to additional self-imposed legal constraints on what the United States is capable of doing in IO. Fortunately, efforts are underway to help resolve this issue, which we will not discuss further in this chapter. The reader should simply be aware that care must be taken to ensure that a given information operation is being performed by the persons or organizations tasked with their assigned role under the U.S. Code in order to keep their activities legal.

Overall, IO by its nature involves a number of complex legal issues. The old legal ways of viewing the use of traditional military capabilities do not provide sufficient guidance for the employment of IO capabilities in the Information Age. Due to the globally connected infrastructure and the anonymity of its participants, there are no longer clear-cut guidelines as to which acts are considered hostile, force, or war. When an action occurs, proving the identity of the perpetrator is difficult at best, and often impossible.

Moreover, the tendency for erring on the side of least risk has led some commanders and planners to avoid options that are, in fact, legal. In at least two recent high-profile cases, poor legal advice by the resident lawyer missed critical opportunities. Improving planning tools to help keep the legal and political considerations in front of the planners and commanders will be essential to good planning and execution monitoring in the Information Age.

References

[1] MITRE News Release No. 658-98, "Joint Task Force on Computer Network Defense Now Operational," McLean, VA: MITRE Corporation and DoD, 30 December 1998; http://www.nap.edu/html/C4I/ch3_b4.html.

[2] The National Security Agency's mission statement from their Web site, http://www.nsa.gov/about/about00003.cfm, March 2006.

[3] The Defense Intelligence Agency's Mission Introduction from their Web site, http://www.diajobs.us/about.html#mission March 2006.

[4] Joint Publication 2-01, *Joint and National Support to Military Operations*, October 7, 2004, http://www.dtic.mil/doctrine/jel/new_pubs/jp2_01.pdf.

[5] Leckie, Robert, *The Wars of America*, Vol. 2, New York: Harper & Row Publishers, 1968.

[6] "Operation Urgent Fury," 1999, http://www.fas.org/man/dod-101/ops/urgent_fury.htm.

[7] Hura, Myron, and Gary McLeod, *Intelligence Support and Mission Planning for Autonomous Precision-Guided Weapons: Implications for Intelligence Support Plan Development*, MR-230-AF, Santa Monica, CA: The Rand Corporation, 1993.

[8] Whitlock, Craig, "Al Qaeda Detainee's Mysterious Release," Washington Post Online, January 30, 2006, http://www.washingtonpost.com/wp-dyn/content/article/2006/01/29/AR2006012901044_pf.html.

[9] *A Baseline Collective Assessment Report for Information Operations*, U.S. Joint Forces Command, December 28, 2001.

[10] Wait, Patience, "DOD Raises the Bar on Info Assurance," February 6, 2006, GCN.com, http://www.gcn.com/25_3/news/38176-1.html.

[11] O'Hara, Colleen, "IT Workers to Receive Pay Hike," FCW.com, November 6, 2000, http://www.fcw.com/article72003-11-05-00-Print.

[12] Vijayan, Jaikumar, "IT Security Destined for the Courtroom," Computerworld.com, May 21, 2001, http://www.computerworld.com/cwi/story/0,1199,NAV47_STO60729,00.html.

[13] Demchak, Chris C., "Fortresses, Badlands, and State Security Paths in a Digital Mass Society," *Cambridge Review of International Affairs Seminar on State Security and the Internet*, London, U.K., May 18–19, 2001.

[14] Arkin, William, and Robert Windrem, "The Other Kosovo War," August 29, 2001, http://www.msnbc.com/news/607032.asp.

[15] Kennedy, Harold, "Marines Bring Iraq Lessons Learned into Street-Fighting Drills," *National Defense Magazine*, February 2006, http://www.nationaldefensemagazine.org/issues/2006/feb/MarineBring.htm.

[16] Allen, Patrick D., *Evolution of Models at the Warrior Preparation Center: Problems and Solutions for Higher-Echelon Exercises*, R-4155-AF/A, RAND, 1993.

[17] Allen, Patrick, "Measures of IO Effects in Experimentation Using the Joint Staff Analysis Model," *Simulation Interoperability Standards Organization (SISO) Conference*, Spring 2002.

[18] "The Joint Experimentation Program—A Laboratory for Transformation," press release, http://www.jfcom.mil/about/experiment.html, November 2005.

[19] DoD Instruction 5000.2, "Operation of the Defense Acquisition System," May 21, 2003, http://www.dtic.mil/whs/directives/corres/html/50002.htm.

[20] Merritt, Ronald, *IO Range Program*, briefing, U.S. Joint Forces Command, November 16, 2005.

[21] *Fundamentals of Naval Weapon Systems*, Chapter 1, U.S. Naval Academy, from the Federation of American Scientists Web site, http://www.fas.org/man/dod-101/navy/docs/fun/part01.htm.

[22] Associated Press, "Hacker Attacks in US Linked to Chinese Military: Researchers," Breitbart.com, December 12, 2005, http://www.breitbart.com/news/2005/12/12/051212224756.jwmkvntb.html.

[23] "An Assessment of International Legal Issues in Information Operations," Department of Defense Office of General Counsel, May 1999, http://www.downloads.securityfocus.com/library/infowar/reports/dodio.pdf.

[24] Ringo, John M., "Get The Lawyers Out of Command," *New York Post*, October 2001, http://www.johnringo.com/fpoplawyer.htm.

[25] Executive Order 12333, United States Intelligence Activities, December 4, 1981; http://www.cia.gov/cia/information/eo12333.html#2.11.

[26] Lotrionte, Catherine, "When to Target Leaders," *The Washington Quarterly*, Summer 2003, http://www.twq.com/03summer/docs/03summer_lotrionte.pdf.

[27] Associated Press, "U.S. Believes Early Saddam Info Accurate," December 15, 2003, http://www.foxnews.com/story/0,2933,105757,00.html.

[28] Associated Press, "9-11 Commission Members Want Atta Claim Probed," CNN.com, August 10, 2005, http://www.cnn.com/2005/POLITICS/08/10/9.11.hijackers.ap/index.html.

[29] *United Nations Manual on the Prevention and Control of Computer-Related Crime*, 1993, pp. 261–264, http://www.uncjin.org/Documents/EighthCongress.html.

Part III
The Future of IO Planning

11

The Future of Information Conflict

> *It is obvious that the media war in this century is one of the strongest methods (of struggle). In fact, its ratio may reach 90 percent of the total preparation for battles.*
> —Osama bin Laden

Information conflict is important now, and as nations' civilian sectors and militaries become more dependent on information technologies, the importance of offensive and defensive IO will continue to increase. On the offensive side, the United States has some advantages, but on the defensive side, it is still vulnerable as we are one of the most information-dependent societies on the planet. Civil cyber-defense will be a key issue for the nation to address in the next decade, or we will face severe consequences.

The real threat to the United States, however, is in the realm of the idea battlespace. The United States is a relative newcomer to the generational conflicts that have been waging there. As a global power, our ideas are contestants in this arena whether we want them to be or not. Our ideas compete with other ideas by their very existence, and there will always be those who consider our ideas as threats to the success of their ideas. Ignoring the threats and new rules in the idea battlespace could potentially lead to long-term national and cultural disasters.

This chapter describes the features of current and future conflict in the idea battlespace, as this is the area where the United States and its allies will need to focus more effort and attention both to address threats in that space, and to advance our ideas in the world arena. This chapter presents the following 15 features of current and future information conflict, as well as recommendations on how to handle them:

1. Everyone has a message.
2. Every idea can rapidly be known globally.
3. Everyone in the world is a potential audience for every message we create.
4. Every group has its own spin on every other group's message.
5. Most messages are *about* a third group, rather than a direct communication between two groups.
6. Initiative matters in the idea battlespace.
7. Accusations are an easy way to gain the initiative.
8. Conflicts in the idea battlespace will often spill over into the physical world.
9. Conflicts in the idea battlespace cannot be won in the physical space.
10. Allies are important in competitions within the idea battlespace.
11. Neither adversaries nor allies are necessarily permanent in the idea battlespace.
12. Some groups will cheat in the idea battlespace and thereby gain an advantage.
13. Multiple dimensions of power are essential to long-term success in the idea battlespace.
14. There is no rear area or sanctuary in the idea battlespace.
15. Conflict in the idea battlespace is persistent, not episodic.

11.1 Feature 1: Everyone Has a Message

Every person or group has an identity: A belief about who we are and what makes us similar to, or different from, everyone else. This identity may be well articulated or latent, radical or calm, clear or confused, consistent or contradictory. For example, the United States has articulated its belief in the benefits of the rule of law, democracy, and free trade. The identity messages of other groups are often not well articulated or consciously thought through. Often, the content of a group's identity message is subconscious rather than articulated until a message appears that helps the group define who they are, or until a message that is in conflict with that group's identity appears.

A group's view of self usually consists of at least three parts: where we are from, who we are, and where we want to be in the future. The differences or conflicts between these three perceptions can lead to motivation for internal change or can cause a group to react against change that they perceive as being forced upon them by others or achieved by others.

For example, in the Middle Ages, the concept of serfdom (who we are) was an improvement over slavery (where we came from), and though serfs had very limited rights, they had a strong desire to maintain those rights. During and after the Black Death, labor became scarce, and the serf class took this opportunity to negotiate for greater freedoms (where we are going), which eventually led to the end of serfdom and the rise of freemen and the middle class [1].

A group's concept of self may also be based on hatred of another group: "We hate this group because they did X to us N centuries ago." Many messages of hate are present throughout the world today [2]. For example, the October 2005 speech of Iranian President Mahmoud Ahmadinejad calling for the destruction of Israel, or the as yet unrecanted messages of Hamas calling for the same, are recent examples of hate messages [3]. However, there are many more hate messages being expounded throughout the world, including within the United States.

Note that in many cases, governments have taken the approach of attempting to send one message to their populace and another to the international arena. The hate messages against Israel spouted by the Iranian President, as mentioned above, are an attempt to build national support and unity, which is playing well to a large part of the Iranian population. At the same time such messages violate the charter of the United Nations, so the Iranian foreign minister has to keep denying such hate messages to the rest of the world—denials that are not meant to be heard by the Iranian people [4]. In this way, the Iranian government is trying to use specific messages to stir up national and religious fervor at home while denying those messages abroad in an attempt to avoid international backlash. Such duplicity in messages is becoming more difficult to achieve and maintain in the Information Age, but governments will continue to employ such techniques until they actually suffer some consequences for such double-speak.

Some messages are more benign than others. Some are purportedly for the benefit of their audience; for example, many religions sincerely believe that proselytizing is in the best interest of the recipient, and that it benefits the group's audience to hear about that group's religion. Conversely, other groups may fear such messages, and try to limit freedom of speech (or freedom of access to messages) specifically to prevent such information from reaching members of their group [5].

Trying to keep track of all these different messages and their effects and interactions is a difficult job, but one that must be accomplished if the United States is to compete in the idea battlespace. We must consider and understand how each of our messages will affect the opinions, beliefs, messages, and even identities of other groups—and how those messages will interact with *other* messages that are being received by the same audience. Otherwise, a U.S. message could unintentionally place the United States at odds with a group we would rather have as a friend.

That is not to say that the objective is for everyone to agree. This is simply not possible. There is no feasible set of conditions that can define a solution that is acceptable to all groups and individuals. There is no "balance point" in the idea battlespace that will satisfy everyone. One reason for this is that there are individuals and groups whose identities and power is derived from conflict, revenge, fear and chaos, rather than from peace and agreement. There will always be groups who view any form of compromise with another group as a loss of identity or power, or who view a peaceful agreement between two other groups as a threat to their interests. It is an academic fantasy that all we have to do is "talk to one another" and we will come to a meeting of the minds. This is simply not true [2].

The objective in planning messages in the idea battlespace is to *understand the implications of each message ahead of time*. Every message will usually include some outcomes that we desire and some that we do not. The objective is to try to ensure that we are not surprised by unintended responses that occur while we are trying to achieve a desired outcome. We need to be able to adequately identify potential undesired outcomes in advance, so that decision-makers can determine whether or not those outcomes are worth risking in order to achieve the desired outcome.

11.2 Feature 2: Every Idea Can Rapidly Be Known Globally

It doesn't take a nation to be on the world stage in the Information Age. Any group, no matter how small, can make its messages known to the rest of the world in the blink of an eye, with practically no investment. The Internet has been called "the great equalizer" because it gives any radical group the same global message power as a world power. It costs practically nothing to obtain a Web address and set up a Web site, and search engines make it possible for like-minded people to quickly and easily find any message. This is one reason why China is so fearful of allowing its citizens free access to the Internet [6].

In addition, interconnections have been increasing among a wide range of other communications devices, and this is only going to increase over time. Cell phone technology now gives access to the Internet as well as vending machines; telephone companies and cable entertainment companies provide Internet access; and television and radio are now reaching broader audiences via the Internet. Another recent addition to global communications is podcasting, which enables groups and individuals to broadcast their message to anyone with a PC, iPod, or other MP3 device. To get noticed on the global scene, all a group needs to do today is have (or steal) a PC, establish a URL, do something shocking (like beheading someone on film), and post it to the Web.

The power of the PC extends beyond mere access. Today, film and photo editing software makes it possible for almost anyone to create high-quality images that *look* real—even if they are not. This makes it much easier to create false messages that audiences will believe, particularly for people who are accustomed to believe that anything they see with their own eyes must be true.

The degree to which information can be passed around the globe also makes it much less easy to keep information hidden. As Robert Redford noted on the thirtieth anniversary of the making of *All the President's Men*, that story could no longer be written today, as the information that the characters were so desperately trying to obtain 30 years ago can now be easily obtained in many ways [7]. This increased openness of information has been beneficial in many ways, not least being the ability to expose nefarious acts that someone is trying to hide. However, it has also made some traditional means of defusing conflicts more difficult.

For example, during the Cuban Missile Crisis, KGB officer Alexander Fomin and a U.S. reporter named John Scali held private meetings to exchange the information they needed to pass without having to go through official channels to see if they could defuse the situation. These informal communications channels were essential to reducing tensions and finding a peaceful solution to the crisis [8]. In today's world, such private meetings are much more difficult to arrange, and the penalties for discovery are greater than ever. For example, when it was announced that some Sunni leaders were meeting with U.S. and Iraqi government officials to seek ways to end the insurgency, some of the Sunni envoys were assassinated [9]. While this does not mean that we should return to the practice of trying to hide information, we must recognize that we will need to find new ways to allow for protected communications between moderates that might otherwise lead to their demise.

11.3 Feature 3: Everyone in the World Is the Potential Audience for Every Message We Create

The United States must assume that every message that it generates will become public, and that even informal or private communications are likely to be exposed. This is not to say that the United States cannot retain military secrets or share them privately with allies, but it does mean that any messages that the United States sends about its intents or interests to a nonally will become public whether the United States intends this or not.

Similarly, public affairs and PSYOP planners must assume that messages aimed at one group *will* be received by another, whether we want that group to receive it or not. For example, a leaflet dropped in a foreign country will be available on the Internet in less than a day; either as an original for sale on eBay or an electronic copy available to download. If not properly prepared, any message that is derogatory toward an individual can be construed or twisted into an attack against the group of which the individual is a member, regardless of the intent. For example, when two members of a U.S. PSYOP unit showed the poor judgment of taunting Taliban fighters to come out and fight while the bodies of two dead Taliban fighters were being cremated, it was caught on film, created an international outcry, and resulted in disciplinary action against them [10].

Moreover, a message that might be sensitive to some allies should be thoroughly explained to those allies before it is made public. Otherwise, our allies will hear the implications of the message from third parties, who have applied their own spin to that message. For example, some of our messages designed to support democracy in the Middle East caused negative reactions in friendly, but not yet democratic, nations [11].

Therefore, we need to have a plan in place that determines who needs to receive what messages and when to ensure that each group understands how that message affects them, and will not be susceptible to having other groups trying to convince them that our message meant something that we did not intend. We need to be proactive in explaining the implications of our messages to the groups that receive them, rather than reactive after others have twisted our messages to suit their purposes. Our enemies will always attempt to manipulate our messages to achieve their objectives, so we need to make sure that we give them no additional ammunition through our own misstatements.

11.4 Feature 4: Every Group Has Its Own Spin on Every Other Group's Message

Every group looks at every other group's messages to see whether they support, threaten, or are orthogonal (irrelevant) to their own messages. When a message is considered a threat, a group will take steps to either discredit the message or the messenger, as described in Chapter 6. In general, a message or messenger can be attacked in one of three ways (note that the term "sender" may refer to an individual or the group from which the message originated):

- By questioning the legitimacy or actions of the sender;
- By questioning the strength or competence of the sender or his or her actions;
- By questioning the friendliness of the sender or his or her actions toward the receiving group.

One means of reducing a group's political power is by questioning its legitimacy, either by claiming that it has usurped its authority or that it was never legitimate in the first place. Whether a group is legitimate or not, most opposition groups will try to make it appear illegitimate in some way so as to reduce its political strength. For example, al-Qaeda declared the provisional government illegitimate because it was not elected. Now that the new Iraqi government has been elected, the claim that it is illegitimate does not carry much weight within Iraq or around the world.

Even if the sending group is considered by most to be legitimate, an opposition group may point to a particular action as being illegitimate, thereby attempting to reduce the sending group's political strength by associating it with an illegal act. The mere accusation of illegitimacy can thus reduce a group's political power, even if the group's actions are perfectly legitimate. For example, whether the U.S. government's wiretapping of foreign calls into the United States is legitimate is still under investigation, but political opponents are using it for near-term political gain to question the legitimacy of the government's actions regardless of whether these actions turn out to be legitimate or not.

Another way to reduce a group's political strength is to question its strength or competence. This approach does not question the legitimacy of the group in terms of how it came into power, but questions the wisdom of whether the group should be allowed to stay in power. If the group has taken actions that can be defined as weak or incompetent, the strength of that group is reduced. Again, all that is required is the accusation of weakness or incompetence, rather than the reality. In the case of the Federal Emergency Management Agency's response to Hurricane Katrina, the government admitted that its response was poor, and this fact is being used to attack the perceived competency of the current administration.

A third way to separate a group from its political power base is to question its "friendliness" toward its constituents or supporters (e.g., by causing other groups to question whether this group is really on their side as it claims to be). If a message can even sow doubt as to a group's friendliness, it will have achieved its goals; again, the accusation of an unfriendly action can carry as much weight, or more, as actual facts. For example, al-Qaeda's efforts to claim the U.S. war against terror is a war against Islam had made some gains in the Muslim world, according to the 2003 Pew Institute survey of global attitudes [12].

In the opinion of this author, the majority of negative messages consist of one or more of these three attack messages. A scan of the media will quickly reveal examples of all three types of discrediting messages. U.S. influence operation planners must be aware of these types of attacks and be prepared to respond. Having contingency plans in place, including branches and sequels, will enhance the speed of response and possibly preempt accusations in these three areas.

11.5 Feature 5: Most Messages Are *About* a Third Group, Rather Than a Direct Communication Between Two Groups

In this stage of the Information Age, the purpose of many messages is not to talk *with* others, but to talk *about* others. This is the era of the third-party message. For example, if Group A (the sending group) wants to gain a political advantage over Group B (the targeted group), it will send a negative message about Group B to Group C (the receiving group). Rather than attempting to communicate with Group B directly, it will simply try to turn Group C against Group B, or gain support from Group C for Group A at the expense of Group B.

For example, the current Iraqi government has been sending the message to the Iraqi populace that foreign terrorists are not friendly toward the Iraqi people and are the real source of instability in Iraq. The government is making no effort to communicate directly with the foreign fighters (not that it would do much good in this case), but is attempting to increase its own political strength by convincing the Iraqi populace that these fighters are bad for the security of Iraq. (Note that in this case the accusation contained in the message is true.) Similarly, the Iranian president's message to his people that Israel should be destroyed is not an attempt to communicate with the government of Israel, but to increase domestic political support for the Iranian government at Israel's expense. (Note that in this case the intended outcome is a violation of the U.N. Charter.)

One way to counter the "U.S. is unfriendly to you" message to groups around the world is to continue to provide humanitarian support around the world when needed. The Pew institute found that in particular the Indonesian population's positive attitude toward the United States improved from 15% to 38% in great part due to our participation in the tsunami relief effort [13]. So when our opponents declare that the United States is "the Great Satan," personal experience to the contrary helps counter that message. The Peace Corps and U.S.-funded nongovernmental agencies also help people realize that Americans care and provide counterexamples to our opponents' messages that the United States likes to kill everyone.

11.6 Feature 6: Initiative Matters in the Idea Battlespace

As with any conflict, the best defense is a good offense, so attaining and retaining the initiative in the idea battlespace are essential to success. Frequently the first message on the subject will frame subsequent discussion and debate. (This is especially true of accusations, which will be described further in Section 11.7.) For example, if someone defines a given situation as being a class struggle, then subsequent arguments tend to focus on why that situation is or is not a class struggle. If someone states that a situation is a matter of racism, then the subsequent discussion is defined by discussing whether or not the situation was racially motivated. The accusation of racism being the cause of slow emergency response in the aftermath of Hurricane Katrina is a perfect example of setting the stage for subsequent debate [14]. All it takes to frame future discussion is to be the first in the arena with a statement.

Unfortunately, the U.S. message development process is so slow and bureaucratic that we are rarely the first to the table with any message. Secretary of Defense Donald Rumsfeld announced in a February 2006 press conference that the U.S. had a "dangerous deficiency" in the area of public affairs, and that al-Qaeda is better at public relations than we are. Quoting directly from the news article [15]:

> The Pentagon chief said today's weapons of war included e-mail, Blackberries, instant messaging, digital cameras and Web logs, or blogs. "Our enemies have skillfully adapted to fighting wars in today's media age, but . . . our country has not adapted," Rumsfeld said. "For the most part, the U.S. government still functions as a 'five and dime' store in an eBay world," Rumsfeld said, referring to old-fashioned U.S. retail stores and the online auction house respectively. U.S. military public affairs officers must learn to anticipate news and respond faster, and good public affairs officers should be rewarded with promotions, he said. The Pentagon's propaganda machine still operates mostly eight hours a day, five or six days a week while the challenges it faces occur 24 hours a day, seven days a week. Rumsfeld called that a "dangerous deficiency."

Just as September 11 was partially blamed on a "failure of imagination," so too are we hampered by our failure to imagine what our enemies can do to us in the idea battlespace [16]. Part of the problem is that, as yet, the standard influence operations planning process does not seem to be sufficiently successful in imagining how the enemy will formulate its messages and distort our messages. Consequently, our opponents are able to define their messages and spin ours in ways we have not yet imagined.

The U.S. needs a better set of processes and tools to enable it to be first with a message, or at least swifter in getting our messages out than in the past. As mentioned in Chapter 8, the use of contingency plans, including branches and sequels, is just as important in influence operations planning as in offensive and defensive planning. We need to anticipate what the enemy will do next and be prepared with our own messages to counter their messages. We must develop ways to attain and retain initiative across the board. Messages to others must be prepared with attention to our allies, so that we can explain our messages to those allies before our opponents impose their own spin upon our messages.

Similar principles apply to those who are not yet allies, including those who are currently our opponents in the idea battlespace.

Information conflicts cannot be won by timidity. We cannot afford to worry about whether everyone "likes us," because we will never be in a situation where everyone will like us. We must accept that as a given. The questions, therefore, are: how do we show decisiveness, strength, legitimacy, and friendliness to those we want to be friendly toward, and how do we get that message out fast and frequently? How do we ingrain the process of looking for threats and opportunities in this abstract space?

Many of the answers to these questions rely on training our influence operations personnel in idea battlespace situation awareness. Our planners need to learn to think broadly about the various viewpoints we need to visualize and consider in our planning, think creatively about what the enemy might do to our messages or about the messages they may promote to which we must respond to, and do both in a timely fashion. This requires extensive education of the influence planning staff; a broad range of expertise to represent the views of so many diverse groups; and the tools and processes necessary to see, plan, decide and monitor our actions in the information space. We will need to ingrain such Red teaming and brainstorming into our message planning process and situation assessment analysis.

11.7 Feature 7: Accusations Are an Easy Way to Gain the Initiative

False accusations are the most common ploy used by the enemy to achieve and retain the initiative in the idea battlespace. Although U.S. law assumes that a party is innocent until proven guilty, most of the rest of the world does not, nor does the media. The assumption of many groups throughout the world (and within the United States) is to believe the worst about the U.S. government and military, and any accusations against the United States simply reinforce what those groups already believe or want to believe. Some audiences naively accept all accusations on the assumption that "if it weren't true, why would they say so?" Critical thinking and considering the source and its motivations are rare commodities in the modern world (particularly if the audience is already preinclined to believe the source of an accusation rather than the target).

During the Korean War, for example, any successful military action against Chinese forces generated a Chinese accusation of war crimes against civilians [17]. Although ludicrously false, such accusations played well to Chinese citizens and allies, and caused the U.S. discomfort in the international community. As mentioned in Chapter 1, Serbs exploited international media to bombard the Allies with a constant stream of accusations related to any collateral damage that might have been caused during bombing [18]. Similar accusations were made during the liberation of Iraq and the battles for Fallujah [19].[1]

Accusations are consistently highlighted by the media, often without question, because accusations represent conflict and controversy. Conflict and controversy sell, and therefore any accusation will have little difficulty attaining media coverage. A good example is the *Newsweek* publication of the accusations that the U.S. desecrated the Koran. *Newsweek* did not sufficiently check the validity of the accusation, but simply accepted the accusation at face value, which led to rioting and deaths overseas [20].

Such is the prejudiced environment in which the United States must compete in the idea battlespace. To help level the playing field, we need to develop additional techniques—such as a quick response team that is prepared for the global information war—for predicting and responding to accusations, and we must be prepared to immediately disprove false accusations. Delays in responding to an accusation can be interpreted either as consent (as in "silence implies consent"), or as resulting from the time required to prepare a lie in response to the accusation. Speed is as important in the idea battlespace as in any other area of conflict, and so long as we fail to be the first to frame future discussions, or to respond quickly and effectively to opposing messages, we continue to forfeit the initiative—and the advantage—to the enemy.

As mentioned in Section 11.6, we need to be proactive in predicting the types of attack messages that will be generated by our enemies. This is likely to require a combination of computer-assisted Red teaming of probable accusations, brainstorming of possible responses, and knowledge-capture techniques that will enable subject matter experts (such as experts on foreign cultures) to codify their knowledge and help us avoid rookie errors. It is essential that we develop the tools, procedures, and rapid authorization to regain and maintain the initiative in the idea battlespace.

11.8 Feature 8: Conflicts in the Idea Battlespace Will Often Spill Over into the Physical World

When a group wishes to gain attention in the idea battlespace, it will often do something in the physical world to gain that attention. As described in Chapter 6, methods of attention-getting range from words to peaceful protests, violent protests, casualties, and deaths, all the way to war. The higher on the scale a group operates, the more attention it is likely to receive from the media. A

---

1. Accusing someone of war crimes uses two attack methods: It questions the legitimacy of the supposed perpetrator, and the friendliness of the supposed perpetrator to audiences who identify with, or who have been told to identify with, the allegedly victimized group.

phrase often used to describe the prioritization scheme for Western news media is "if it bleeds, it leads." This is one reason why antiglobalization protests typically involve the destruction of property: Such attention-getting actions ensure front-page coverage in the media.

Even without the incentive of media attention, however, there are other reasons for conflicts in the idea battlespace to spill over into the physical world. Frustration is one of the most common triggers, as individuals become fed up and others follow suit. The 1965 Watts riots are an example of such an event, as were the widespread car burnings in France in 2005. In the international arena, frustration with the United Nations' inability to enforce its resolutions, no matter how blatant the violations, was a contributing factor to the Coalition invasion of Iraq.

There will always be disaffected groups, both within the United States and abroad, who will use violence to advance their agenda. When this occurs, the United States must avoid the all-or-nothing thinking that assumes that the United States is somehow at fault or that we "caused it to happen." The use of violence has always been a common choice for those who want to gain attention for their message, and simply talking to them or trying to reach a meeting of the minds will not solve the problem or remove the incentive to use violence. As mentioned earlier, some groups gain and retain power only through violence and hate messages and have no interest in peace, as in many cases peace would result in the loss of the group's power or even its very existence. The United States needs to better prepare the expectations of its people as to who is likely to choose violence over peace and why, so that when that inevitable violence occurs, it does not come as a shock to the American people.

11.9 Feature 9: Conflicts in the Idea Battlespace Cannot Be Won in the Physical Space

Although conflicts in the physical space can be triggered in the idea battlespace, the ultimate conflict cannot be resolved in the physical space. While steps can be taken to reduce the violence, the choice to use violence must be addressed primarily in the idea battlespace. All conflicts in the idea battlespace need to be resolved in the idea battlespace, because that is where the cause is located. Resulting activities, such as violence, in the physical space are only symptoms of the cause, and treating the symptoms alone will not resolve the causes.

For example, the U.S. Army realizes that in Iraq, as in all insurgencies, we cannot win through military might alone. Such a conflict would drag on indefinitely, or until U.S. political will gave out. Thus, some of our allies are making progress in defeating the ideology behind terrorism. Since killing innocent civilians, especially women and children, is against the Koran, a number of Imams have begun showing terrorists the error of their ways and turning them against their former associates. For example, several Jordanian Imams have succeeded in convincing former terrorists that their actions are against the teachings of the Koran and that those who taught them to act against the Koran were false teachers [21]. In a similar manner, when tribe members kidnapped Westerners in Yemen, tribal leaders convinced the perpetrators that their actions threatened the well-being of the tribe. The fact that the tribal leaders also brought in the Yemeni military to surround the kidnappers also helped the negotiations [22]!

When foreign fighters began to enter Iraq, they were often welcomed by the insurgency as bringing capabilities and skills to fight the Coalition forces. Once the foreign fighters began killing Iraqis as examples or as victims of frequent terrorist bombings, however, their appeal diminished significantly among the largely Sunni insurgency. In addition, the capture of the message from Ayman al-Zawahiri, al-Qaeda's number-two leader, to al-Zarqawi made it clear that the local insurgency was simply being used by al-Qaeda and would be relegated to a minor role once al-Qaeda had won in Iraq [23]. As a result, a number of foreign fighters in Iraq have been killed by the insurgents themselves [24]. More importantly, the insurgency and in particular the foreign fighters have become less and less popular with the Iraqi population as a whole, as evidenced by the many tips coming from the anonymous tip line in Iraq, which has led to a number of terrorists being captured and terrorist missions being thwarted [25].

Actions in the physical world are often deemed necessary for groups to gain attention and are equally necessary for other groups to control the situation. One cannot win the peace by weapons alone, however; true success or failure can only be achieved by winning in the idea battlespace, not in the physical world. Thomas Hammes describes this type of conflict as "4th Generation Warfare" [26].

11.10 Feature 10: Allies Are Important in Competitions Within the Idea Battlespace

In the idea battlespace, it is difficult for any group to win alone. One of the symbols of power in ancient Rome was the fasces or bundle of sticks, which represented the strength of the group compared to the weakness of the individual. (The same symbol was used for a time on the U.S. Mercury dime.) The strength of the group has been recognized in the formation of nations, and of international communities of interest such as the United Nations. In the idea battlespace, the voice of such a group carries more weight than that of an individual (such as a single nation) who opposes the group's ideas.

This is not to say that groups are always right and individuals are always wrong. On the contrary, there are many examples of just the opposite. However, when it comes to political power, when more than one group is in agreement, this tends to carry more weight than the voice of a single group. Thus, if the United Nations passes a resolution that Iraq or Iran must allow nuclear inspectors, this carries more weight on the international scene than such a demand from a single nation, such as the United States. (Of course, some nations have figured out that such resolutions without enforcement involve no consequences and thus ignore the voice of the group [27].)

Similarly, when the United States has worked with our allies to present combined messages, these messages have carried much more weight than messages that the United States has sent on our behalf alone. When we have spoken or acted alone, it has tended to cost us in the international arena. This is not to say that the United States should allow itself to be incapacitated by being unable or unwilling to act without the support of our allies, but we must also be aware of, and prepared for the consequences of, marching off alone against the wishes of those allies. While we will never satisfy every ally, we need to be aware of the consequences of acting against the wishes of the larger group and take steps to mitigate those effects before we do. Before taking an action, we need to make our explanation and rationale as clear and convincing to as many allies and neutrals as possible, as well as to our enemies themselves. We must also keep in mind that those who are allies today may be opposed to us tomorrow, and vice versa—and make sure that our messages do not burn any bridges.

11.11 Feature 11: Neither Adversaries Nor Allies Are Necessarily Permanent in the Idea Battlespace

Over time, those who agree with us now may disagree with us later, while those who oppose us now may agree with us later. Therefore, it is not a good idea to burn bridges with adversaries who may become future allies, or to rely on or take for granted current allies who may later oppose us. As Prime Minister Disraeli once said, nations have no permanent enemies or allies—only permanent interests [28]. This has always been true on the domestic and international scene and applies equally in the idea battlespace.

Many in the United States need to get past the all-or-nothing thinking about friends and enemies. Allegiances shift over time, especially in tribal-based cultures where loyalties are based more on kinship or "what can you do for me now?" than on ideology. For example, bin Laden and his followers reportedly bribed some of our Afghan allies to allow him to escape from the Tora Bora trap [29]. Conversely, the United States has been helping the new Afghan government use carrot-and-stick approaches to bring some of the Afghani warlords and their militias into compliance with the government. One of the lessons learned from the Afghanistan stabilization effort is that there frequently are not any clearly "good guys" (at least by U.S. standards), and that the "bad guys" don't always remain bad [30].

This is not to say that the United States should compromise its principles, but it must understand that people who need to perform their own nation-building or develop group identities typically have very different principles, often aimed at self-interests or tribal interests. Even in Iraq, some Sunni insurgents who vowed to fight the U.S. "occupiers" to the bitter end have realized that the foreign terrorists were far more dangerous to them, and that the United States is seeking to encourage Shi'ite compromises rather than complete Shi'ite political domination. In short, the United States is beginning to look like one of the few allies the Sunnis have in Iraq at the moment, and they may not wish to keep opposing every U.S. effort [9].

In the stabilization scenarios we have encountered, allegiances of the various groups and parties shift widely and often. Our analytic and other decision support tools need to be better able to address these shifting interests. Planners must be aware that there are not simply two sides to a conflict—those who are shooting at us versus those who are not—because members of each side may do one on one day and the other on another. We need analytic tools that better describe the groups we are dealing with in every locale in a stabilization operation. Some of the questions these tools need to address include:

- Who gets along with whom?
- Who is often fighting whom?
- Who has the power, and which types of power do they wield?
- Who are the "disruptors," the disaffected who have nothing to lose and who gain power only through chaos?
- Who are the local and/or traditional providers of food, water, medicine, and security?
- How do we operate in coordination with them while giving them the credit?
- How do we get armed groups to disarm, disperse, or reorganize into a more constructive organization?
- What are the cultural sensitivities of the locals, and how do you convey these clearly and succinctly to the commander and troops on the ground?

We need better planning, analytic, and other decision support tools to help win the peace both before and after the conflict. We need to represent the true interests of these various groups, without rigidly defining them as contrary to or in line with our interests. Sometimes, "good enough" stabilization is the best we are going to get, but even that will not be achieved if we regard a group as once an enemy, always an enemy. If our tools and analytic techniques can include accurate representations of group interests, we have a better chance of finding common ground with all but the most die-hard opponents.

11.12 Feature 12: Some Groups Will Cheat in the Idea Battlespace and Thereby Gain an Advantage

While the U.S. populace, media, and government all claim to value the truth, we must be aware that this is not true of every country or group in the world. Many groups are playing for keeps, and in their minds, the end justifies the means. Lying is simply one tool to achieve one's goals, so to such a group, the truth is valuable only if it supports the group's position, and lying is easily justified if it supports their goals. For many nations, truth is regarded as a liability rather than a virtue, while lying provides an advantage (at least temporary) over the populations toward which the lies are aimed. Since so many of our opponents believe that the end justifies the means, including intentionally targeting innocent civilians and killing prisoners, lying is simply another tool in their arsenal.

Most opponents of the United States lie as part of a strategy to counter U.S. strengths. Since no opponent can defeat the United States in open battle on land, sea, or air, the opponents must try to defeat us in the idea battlespace. For example, during the Korean War, the Chinese used all sorts of lies to bolster support at home and degrade international and U.N. support for the war. In addition to the trumped-up charges of atrocities described earlier, the Chinese used numerous lies during the peace talks. Thus, when the U.N. peace talk delegation arrived in Chinese-controlled territory, the Chinese media made it appear that the U.N. was beaten and suing for peace, since the only people with weapons in the pictures were Chinese, and the cars of the U.N. visitors flew white flags. As another example, they propped up the chair of the diminutive Chinese representative and shortened the chairs of the U.N. delegates, so that the Chinese would appear to be the domineering side during the talks. These photographs were a propaganda coup, as they played well to China's population and its allies [31].

Instruments of al-Qaeda propaganda frequently lie about U.S. atrocities against innocent Muslims, thereby generating volunteers for the Islamic jihad. The message is always that the United States is intentionally targeting innocent civilians, despite the fact that the atrocities never occurred and in most cases the supposed live film footage is file footage [19]. In other cases, events were "recreated" by Middle Eastern reporters to try and drum up outrage. The events may or may not have occurred as they were portrayed, but to represent a recreation as a film of the actual event is clearly a lie [32]. Truth, however, means little to those who believe they are partaking in a holy crusade for Islam against the West [2].

Another example is the claim of Iranian President Mahmoud Ahmadinejad that the 2006 damage to the Shi'ite mosque was performed by the United States or the Israelis [33]. The mosque was actually damaged by terrorists who were trying to cause a civil war in Iraq, an act which caused outrage among the mostly Shi'ite population. However, since the Iranian government did not want to blame the actual perpetrators (terrorists fighting the United States and the new Iraqi government), but needed to assuage the primarily Shi'ite populace in Iran, the Iranian government claimed that the crime was committed by the "Zionist" United States and Israelis. Though the claim was blatantly ludicrous, it played well in Iran. Populations that are not trained in critical thinking are perfect targets for such lies, especially when the population has been prepped with a constant barrage of hate messages.

When the United States is fortunate, sometimes the enemy's lies are so obvious as to be beneficial to the United States. Baghdad Bob is a perfect example. The pronouncements of Mohammed Saeed Al-Sahhaf, the Information Minister of Iraq during the U.S. invasion, included incredible claims of Coalition defeats. One of the more incredible was that [34]: "One hundred infidels committed suicide as they entered the holy city of Baghdad." His outrageous claims were so useful to the United States that even President Bush said of Baghdad Bob [34], "He's my man, he was great. Somebody accused us of hiring him and putting him there. He was a classic." The United States is not usually so fortunate as to have such an incompetent liar opposing us, however. Therefore, we need to be prepared to counter the lies of skilled and practiced professionals.

Just as U.S. influence planners need to be prepared to address the spin that opposing groups put on our messages and to handle false accusations, we must also be better able to predict an enemy's lies and have preplanned responses prepared to quickly demonstrate that they *are* lies. For example, as described in Chapter 8, having footage over time of the supposedly bomb-damaged mosque during Operation Desert Storm was an example of having proof that the damage was inflicted by the Iraqis [35]. We need to be better prepared for future events so that we can counter lies and have a ready response. There may even be opportunities to "poison the well" of certain lies by predicting the types of lies that are likely to be forthcoming and taking preemptive action with our own messages.[2]

---

2. Note that U.S. law prohibits the U.S. military from using PSYOP against the American public. What we are talking about here is using influence operations to tell the truth and expose enemy lies, not to perpetrate lies.

11.13 Feature 13: Multiple Dimensions of Power Are Essential to the Long-Term Success in the Idea Battlespace

So our enemies lie. This is nothing new. Throughout the last century, our opponents have been liars more often than not. We still beat them in most cases. What is different in today's world is the global reach of those lies, which makes them more effective than ever before.

Since the opponents of the United States are weaker than the United States in economic and military capabilities, they attempt to exploit informational and diplomatic means to gain power and influence to achieve long-term objectives. They do not espouse truth, fair play, or protection of innocents as values, while the United States does. This does give the terrorists an advantage in several ways, but the answer is not for the United States to drop our values. The United States has often fought opponents who have lied, killed civilians, and/or used human shields, from Germany and Japan in World War II, China and North Korea in the Korean War, North Vietnam and the Viet Cong, and Saddam's Iraq. In most of these cases, the United States was able to overcome the advantages gained by opponents who have not played by the same rules and constraints by which we live.

Our previous success, however, has been based on the richness or strength of our nation across all dimensions of diplomatic, information, military, and economic instruments of national power. A true superpower is strong in all of these areas, which is why we have previously been able to withstand situations in which our opponents have had the advantage in one or two of these other areas. Therefore, if the United States applies its resources behind our information instrument of national power in a much more coherent and determined fashion, we will be able to defeat our opponents in the Information Age as well.

The advantage of being a superpower is that we do have great depth in all of the dimensions necessary to win this conflict, even with our self-imposed constraints. But we need to organize and equip ourselves for dealing with a domain that is likely to dominate all future conflicts for at least the next century—the information sphere.

We also need to avoid tolerating double standards. For example, when Communism was popular, repressive Communist government atrocities were generally ignored by the U.S media while those committed by any U.S. ally were broadcast around the world. While President Jimmy Carter's emphasis on human rights was a brilliant move, it was not uniformly applied to all nations until Ronald Reagan declared the Soviet Union to be an "evil empire." This was the first effort to hold the Soviet Union to the same standards as the rest of the world and was successful, contributing to the downfall of the Soviet empire.

The United States needs to do the same in the war against terror. Tolerate no duplicity that condones acts of terrorism. Speak out when such hypocrisy is observed. Allow no sanctuary to those that would hide behind a double standard. As mentioned earlier, the conflict in the idea battlespace is no place for the timid. If something is important, it will be controversial.

11.14 Feature 14: There Is No Rear Area or Sanctuary in the Idea Battlespace

In previous wars, the United States usually benefited by having rear areas or sanctuaries where our populace could be protected. During the Cold War, the threat of a nuclear holocaust hung over the nation, but except for the Cuban missile crisis, tensions between the United States and the former U.S.S.R. did not rise to significant levels. Sanctuary, however, is no longer the case for the U.S. population in the Information Age. Not only is there the physical threat of terrorism on U.S. shores, but also the Information Age allows all competing messages from the idea battlespace to reach the American population. From racist hate messages to "the United States can do no right" messages, all sectors of our society are being reached. As a result, the U.S. population is constantly being bombarded by lies and false accusations, which wear on the U.S. population over time and erode support for any U.S. endeavor.

What should the United States do about protecting the U.S. population from lies and false accusations? The answer is *not* censorship. The Information Age has precluded the censorship option in all but the most oppressed nations, such as North Korea and China. (Even then, China is having great difficulty maintaining a lid on free access to information when they want free access to markets.)

The answer is in placing emphasis on educating people how to think critically for themselves. When using the term "education," the author is *not* talking about re-education camps, mind-programming, brain washing, or an Orwellian *1984* education, but rather about educating the U.S. population in critical thinking. The nation needs to emphasize courses in critical thinking, not in political correctness or party lines (for either party). In such classes, we need to provide examples of lies and how they can be identified and countered. We need to explain how people can be readily manipulated if they do not think critically. We need to help our citizens become better able to survive and thrive in the Information Age.

Such teaching of critical thinking was supposed to be happening in our schools and universities, but that is no longer the case [36]: "Most undergraduates leave college still inclined to approach unstructured 'real life' problems with a form of primitive relativism, believing that there are no firm grounds for preferring one conclusion over another." In a similar manner, Mortimer Adler described why students are no longer being effectively taught critical thinking [37]:

> Those engaged in educationally profitable discussion will be engaged in agreeing or disagreeing, arguing when they disagree, and giving reasons for disagreements. They will be making and defending generalizations, or challenging generalizations made by others. They will be judging by weighing evidence pro and con, or by examining the validity of reasons for making one claim or another concerning what is true or false, more or less probable. They will be asking and answering questions about the consistency or inconsistency of things asserted or denied, about their presuppositions and their implications, and about the inferences involved therein. . . . How are such intellectual habits of skill developed? Exactly in the same way that all bodily habits of skill are developed: by coaching, not by didactic instruction using textbooks that state the rules to be followed. . . . The programs in critical thinking now being advocated from coast to coast are minuscule and oversimplified versions of the much more rigorous course in logic that I taught in college. And they will be just as ineffective in producing students who can think critically in other courses. Nor will they train teachers how to think critically. That training should have been accomplished by the education they received before they started to teach.

One of the most telling experiments recently performed was brain scans of people self-identified as being part of a given party, left or right. The subjects were exposed to contradictory information about their party, and the scans showed that the part of their brain associated with critical thinking was literally shut off when exposed to such contradictions. Neither party had a monopoly on mindlessness [38]. This experiment goes a long way in expressing the degree to which the polarization of the nation has reached and how far we need to go to teach people how to think critically again.

To help the nation regain its powers of critical thinking, it must first recognize that the Information Age has, by its nature, made it easier to gain political power by talking *about* others rather than with them. We must, individually and collectively, decide to leave the comfort of being mentally unchallenged and address topics that make people feel uncomfortable. And that means that citizens from both ends of the political spectrum need to examine the lies and hate messages they have accepted simply to support one party or the other.

Hate messages must be exposed, whether they are hate messages against a race, a religion, a political party, or unpopular beliefs. Group-level all-or-nothing thinking along the lines of "someone who disagrees with me is evil and anyone who agrees with me is good" is not how this nation was built or how it has survived to date.[3]

After we educate the U.S. population in critical thinking, we need to make similar efforts abroad. Not every nation will welcome such efforts, but some will. Many nations send their students to the United States to learn, and these are good opportunities to teach critical thinking. U.S. citizens travel and work abroad, and these are other opportunities to share how to think critically.

Take the example of women's rights. The United States supports enhancing women's rights around the globe, but many cultures oppose it. The United States as a nation is trying to find ways to enhance women's rights without alienating a few of our allies who are stuck in the Middle Ages in that area. Much of the opposition, however, comes from groups within the United States who support the position that all cultures should "be themselves" even at the expense of oppressing their women. These two competing concepts are mutually contradictory, and both cannot eventually win in the idea battlespace. Yet the debate continues to rage even in this country, with some groups claiming to support both contradictory positions simultaneously. Such logical inconsistencies in our messages must be exposed for what they are.

Consistency in the U.S. position on any issue will be important in the Information Age. Karen Hughes, Undersecretary for Public Diplomacy, is helping the United States present a more coherent message by stating the themes and messages the ambassadors should present. This is a step in the right direction.

Besides educating Americans in critical thinking, we also need to do a better job of explaining who the United States is both to our own citizens and to the rest of the world. We need to *articulate* who we are and what we are about, to ourselves and to others. We must explain that the reason we bicker among ourselves is not a sign of weakness, but of long-term strength, whereby we constantly question our own motives and actions to make sure we are doing the right thing. When we make a mistake, we must admit it to everyone, not just to ourselves. There are many aspects to American life and culture that are desired around the world, things which many of us take for granted. We need to keep at the forefront of our strategic communications the message of who we are, why we are, and what it means to the rest of the world, and then make sure our actions back up our words.

To summarize, the way to protect U.S. citizens in the Information Age is not a matter of censorship of lies and false accusations, but of inoculating U.S.

---

3. Note that toleration does not imply or require agreement. Toleration does not mean that I have to agree with you to avoid offending you or hurting your feelings. Toleration means that I do not kill you because I disagree with you. Toleration must also be reciprocal, or it is simply coercion of one group by another.

citizens by teaching them how to think critically and how to identify and counter lies and false accusations. Armed with descriptions of common techniques in lying, the American and other populations will be better able to evaluate the lies in the messages coming from our enemies and even from within. We need to help citizens of the United States and much of the rest of the world identify the lies and recognize them and then come to their own conclusions. We also need to be clear in our messages and our actions as to who we are and what we stand for, so that everyone clearly understands what we are doing and why—whether they agree with it or not.

11.15 Feature 15: Conflict in the Idea Battlespace Is Persistent, Not Episodic

There are no treaties, truces, or time-outs in the idea battlespace. Unlike physical conflict, which can be episodic, conflict in the idea battlespace is constant and persistent. The conflict among ideas goes on 24 hours a day, 7 days a week, and lasts for generations.

We must plan, organize, resource, and train for this *persistent conflict* in the idea battlespace, or face the consequences of our lax attitude. Donald Rumsfeld, as quoted in Section 11.6, recognizes that we are not yet operating in a 24/7 mode in the public relations or strategic communications realm, and we need to be. It is ridiculous for a superpower to have not yet assigned the resources (people, processes, and funding) to compete with an opponent in the idea battlespace. Since bin Laden himself has stated that the idea battlespace is where 90% of the battle will be in this age, we need to start allocating our resources to where the battle is and will be, rather than just to where we are already good at fighting and winning.

Part of the problem the United States faces in adapting to the Information Age is that we are not desperate enough. We are not yet willing to make some of the drastic changes necessary to adapt to the new age. For a short while, we were hurting enough. September 11 caused the most significant upheaval in how the United States government operated since Pearl Harbor. Our military operated differently than ever before in Afghanistan and won quickly in one of the most remote places on Earth where the Soviets had failed after 10 years of trying. We then invaded Iraq and won so quickly that many were surprised by the sudden collapse of resistance [39]. Once again, we did it differently, causing the surrender of large populated areas faster than any time in history. As one senior British officer observed [40], "Forget everything that you have read in the history books about modern combat; this is going to be different." It was different because we adapted to the new environment more quickly than our opponent. We then learned what we did wrong in trying to win the peace in Iraq and continued to adapt to that new environment [41].

A similar adaptable mentality is required to succeed in the idea battlespace. We need a 24/7 strategic communications capability with sufficient authority to develop, approve, and disseminate messages in rapid time to obtain the initiative and defeat our opponents. We need to organize a national capability to wield the information instrument of national power to coordinate and dominate the persistent conflict in the idea battlespace. We need to allocate all of the supporting capabilities necessary to make it happen, as will be described in Chapter 12. We need to develop the tools and techniques that support planning and operations that adequately address the 15 features of conflict in the Information Age. We need to be proactive in predicting the types of attacks we will face in the idea battlespace. We need feedback loops to ensure we can adapt our planning factors to address when they are no longer working and why.

These 15 features define current and future conflict in the idea battlespace. We have recommended some concepts and techniques to address each one of them. The recommendations provide a mutually supporting set of capabilities to compete and win in the idea battlespace against opponents who lie, allies and enemies that change allegiances, a politically polarized nation, and media often hostile to any government or military initiative, and groups on the international scene who have been taught to hate us for generations. Were the United States not a superpower, the task would be daunting. But even this superpower needs to marshal all its resources and compete to win in the idea battlespace, where we have been taking a beating for decades. It is time to get serious about competing in the idea battlespace and allocate the resources necessary to succeed.

References

[1] Shama, Simon, *A History of Britain*, New York: Hyperion, 2000, pp. 236–237.

[2] Peters, Ralph, "The Shape of Wars to Come," *Armchair General*, March 2006, pp. 68–75.

[3] "Mahmoud Ahmadinejad," May 22, 2006, http://en.wikipedia.org/wiki/Mahmoud_Ahmadinejad.

[4] "Tehran Says Not Threatening Attack on Israel," MSNBC.com, October 29, 2005, http://www.msnbc.msn.com/id/9823624/.

[5] Strode, Tom, "Lack of Afghan Liberty May Threaten Aid," *Faith & Family News,* January 31, 2003, http://sites.silaspartners.com/partner/Article_Display_Page/0,,PTID314166%7CCHID605964%7CCIID1569564,00.html.

[6] Associated Press, "Tech Giants-Lawmakers Debate Censorship," February 15, 2006, http://www.cnn.com/2006/TECH/internet/02/15/us.web.censorship.ap/index.html.

[7] "'No One Wants to Hear About This:' Robert Redford Looks Back on *All the President's Men,*" CNN.com, February 21, 2006, http://www.cnn.com/2006/SHOWBIZ/Movies/02/21/film.allthepresident.ap/index.html.

[8]   Watson, Mary Ann, *The Expanding Vista,* New York City: Oxford University Press, 1990, excerpt on the Cuban Missile Crisis obtained from the Museum of Television and Radio Web site, http://www.mtr.org/events/satellite/cuba/cuba1.htm.

[9]   Johnson, Scott, Rod Nordland, and Ranya Kadri, "Exclusive: Direct Talks—U.S. Officials and Iraqi Insurgents," *Newsweek,* reported on MSNBC online news, February 6, 2006, http://www.msnbc.msn.com/id/11079548/site/newsweek/.

[10] "U.S. Soldiers Reprimanded for Burning Bodies," CNN.com, November 26, 2005, http://www.cnn.com/2005/WORLD/asiapcf/11/26/afghan.us.soldiers/index.html.

[11] Dickey, Christopher, "Battleground of Ideas," *Newsweek* Web exclusive commentary, February 1, 2006, http://www.msnbc.msn.com/id/11128153/site/newsweek/.

[12] Pew Institute, "War with Iraq Further Divides Global Politics," *Views of a Changing World 2003,* report released June 3, 2003, http://people-press.org/reports/display.php3?ReportID=185.

[13] Pew Institute, "U.S. Image Up Slightly, But Still Negative," *Pew Global Attitudes Project,* report released June 23, 2005, http://pewglobal.org/reports/display.php?ReportID=247.

[14] "Katrina Victims Blame Racism for Slow Aid," MSNBC.com, December 6, 2005, http://www.msnbc.msn.com/id/10354221/.

[15] "Rumsfeld: Al Qaeda Has Better PR," CNN, February 17 2006, http://www.cnn.com/2006/POLITICS/02/17/security.rumsfeld.reut/index.html.

[16] Grier, Peter, and Faye Bowers, "Failure of 'Imagination' Led to 9/11," *Christian Science Monitor,* July 23, 2004, http://www.csmonitor.com/2004/0723/p01s03-uspo.html.

[17] Middleton, Pete, "When Chinese Troops Fired on Two Gloster Meteors at Chongdan, the Australians Made Them Regret It," *Military History,* August 2005.

[18] Thomas, Timothy, L., "Kosovo and the Current Myth of Information Superiority," *Parameters,* Spring 2000, http://www.carlisle.army.mil/usawc/parameters/00spring/contents.htm.

[19] Peters, Ralph, "Kill Faster!" *New York Post,* May 20, 2004.

[20] Van Zandt, Clint, "Media: Irresponsible, Stupid, or Calculating?" MSNBC.com, May 19, 2005, http://www.msnbc.msn.com/id/7910327/.

[21] Phares, Walid, "11/9," *Front Page Magazine,* November 10, 2005, http://www.frontpagemag.com/Articles/Readarticle.asp?ID=20140.

[22] "Freed Italy Hostages Tell of Ordeal," CNN.com, January 6, 2006, http://www.cnn.com/2006/WORLD/meast/01/06/yemen.kidnappings.ap/index.html.

[23] Zawahiri, Ayman, alleged letter to al Zarqawi, http://www.dni.gov/.

[24] Gilmore, Gerry J., "Iraqi Insurgents Now Battling Al Qaeda Terrorists," *American Forces Press Service,* January 26, 2006, http://www.defenselink.mil/srch/docView?c=A3B245203F9EBEC5FC8C306E4AAF152F&dk=http%3A%2F%2Fwww.defenselink.mil%2Fnews%2FJan2006%2F20060126_4023.html&q=terrorists+%3Cand%3E+insurgents+%3Cand%3E+battling+%3Cand%3E+qaeda+%3Cand%3E+iraqi+%3Cand%3E+now+%3Cand%3E+al&p=Simple.

[25] Behnam, Babak, "Getting Tips on Iraq's Most Wanted," MSNBC.com, April 26, 2005, http://www.msnbc.msn.com/id/7641117/.

[26] Hammes, Thomas X., "Insurgency: Modern Warfare Evolves into a Fourth Generation," *Strategic Forum*, January 1, 2005.

[27] "The U.N. & Iran: Weak Saber-Rattling," *Pittsburgh Tribune-Review* editorial, April 7, 2006, http://www.pittsburghlive.com/x/tribune-review/opinion/archive/s_440890.html.

[28] Janes, Jackson, "Benchmarking as a Tool for Alliance Management," the Globalist Web site, August 6, 2003, http://www.theglobalist.com/DBWeb/StoryId.aspx?StoryId=3344.

[29] Smucker, Philip, "How Bin Laden Got Away: A Day-by-Day Account of How Osama Bin Laden Eluded the World's Most Powerful Military Machine," *The Christian Science Monitor*, March 4, 2002, http://www.csmonitor.com/2002/0304/p01s03-wosc.html.

[30] Maloney, Sean M., "Afghanistan Four Years on: An Assessment," *Parameters*, U.S. Army War College, Autumn 2005.

[31] Hermes, Walter G., *Truce Tent and Fighting Front*, Center of Military History, 1966, http://www.army.mil/cmh-pg/books/korea/truce/fm.htm.

[32] Goldberg, Vicki, "Seeing Isn't Believing: When Pictures Become Propaganda, History Can Take a Wrong Turn," *Readers' Digest*, September 2004, pp. 142–146.

[33] "Iran Leader Blames U.S., Israel for Mosque Blast," MSNBC.com, February 23, 2006, http://www.msnbc.msn.com/id/11515193/.

[34] Al-Sahhaf, Mohammed Saeed, "Iraqi Information Minister Quotes," http://www.militaryquotes.com/, April 2005.

[35] Parks, W. Hays, "The 1954 Hague Convention for the Protection of Cultural Property in the Event of Armed Conflict," http://www.kakrigi.net/manu/ceip4.htm.

[36] Bok, Derek, "Are Colleges Failing? Higher Ed Needs New Lesson Plans," *Boston Globe*, December 18, 2005, http://www.boston.com/news/education/higher/articles/2005/12/18/are_colleges_failing?mode=PF.

[37] Adler, Mortimer, "Critical Thinking Programs: Why They Won't Work," archived on the Radical Academy, http://radicalacademy.com/adlercritthinkingpro.htm.

[38] Bourg, Jim, "Political Bias Affects Brain Activity, Study Finds," Reuters, January 24, 2006, http://www.msnbc.msn.com/id/11009379/.

[39] Fox News Staff, "Arabs Shocked, Awed by Fall of Baghdad," Fox News.com, April 9, 2003; http://www.foxnews.com/story/0,2933,83704,00.html.

[40] Johnson, Marcus, "Iraq: A New Kind of War," BBC Online News, April 14, 2003, http://newswww.bbc.net.uk/2/hi/middle_east/2946597.stm.

[41] Dunn, J. R., "Prospects of Terror: An Inquiry into Jihadi Alternatives Part I," The American Thinker Web site, March 21, 2006, http://www.americanthinker.com/articles_print.php?article_id=5345.

12

Planning for Success in a New Age

The future ain't what it used to be.
—Yogi Berra

The world has entered the Information Age, and most of the world is still trying to adapt to this new age. What are the new rules? How do things work? What does not work the way it used to? How can one gain an advantage? What are the threats? What are the opportunities? How does one organize, prepare, and plan for the Information Age?

Sometimes the answers to these questions will be correct, and sometimes they will not be. In most cases in which the answers are not correct, it is because the observer is stuck in old ways of thinking. As described in the books *Who Moved My Cheese?* [1] and *The Innovator's Dilemma* [2], those most susceptible to continuing in the old ways after the rules have changed are those who were successful in the old age. Those who tend to be most successful in adapting to a new age are those who were not yet successful in the previous age and need to be successful in the new one.

Section 12.1 briefly explores how to recognize when you are in a new age, describes some of the key features that have changed in the Information Age, and presents some of the problems and opportunities of being in the Information Age. Section 12.2 makes a recommendation for a significant change to support how we should organize and operate in the Information Age: by defining the information sphere as its own domain similar to air, land, sea, and outer space, and then developing a combined Military and Interagency Service to operate effectively in this new domain.

12.1 Revolutions in Military and Political Thought

Revolutionary ideas appear periodically over time. Someone perceives that the rules of the game have changed and defines a new paradigm or mental model for how to succeed in this new environment. These new concepts or mental models change our views of the world, which in turn lead to new capabilities that did not previously exist, or were never used together before in ways that led to clear advantages on the battlefield or the political landscape.

12.1.1 Recognizing That We Are in a New Age

The first prerequisite for defining a new age is to recognize that you are in one. Gustavus Adolphus recognized that gunpowder-based firepower was becoming superior to the melee power of pikes. Napoleon and other leaders of the First Republic recognized the power of the mobilizing a nation for increased military manpower through national fervor. Douhet and Mitchell recognized the power and effectiveness of airpower before the Army or Navy Military Services did. Rommel and Guderian recognized the need for protected mobility to overcome static defenses, and put together technologies and techniques not previously combined. Nuclear weapons changed the concept of winning into a situation in which there was a good chance that no one could win.

Each military revolution was preceded by a period of invention, and the revolution occurred when someone put all the necessary pieces together in an innovative way—a way that no one had considered before. For example, Gustavus Adolphus put together the light artillery, prewrapped cartridges, the shock value of saber-wielding cavalry, and innovative logistics for his day. (He also emphasized cannons on ships rather than simply using ships as melee platforms.) Along with the use of the indirect approach, he revolutionized land warfare from primarily a melee conflict to a conflict dominated by firepower and maneuver.

The U.S. Civil War established the superiority of accurate fires and prepared defenses over massed formations in the open, the advantages of rapid mobility of rail and water transportation, the power of ironclad ships, the effectiveness of the blockade, and the value of striking at the opponent's logistics infrastructure to reduce his war-making capabilities.

The Germans took the infantry storm tactics proven in World War I, along with new armor, communications (a radio for every tank platoon), and close support aircraft, and turned it into the revolutionary warfare of its day: the blitzkrieg. They may also have read Liddell Hart's articles on strategy and the emphasis on the indirect approach, which forces the enemy to make decisions when already on the horns of a dilemma [3, 4].

Each of these military revolutions took recent technological and tactical developments and combined them in ways that had not been seen before, contributing greatly to their success on the battlefield. The key is not so much to win the technology race but to combine new technologies and techniques with new ways of thinking, or new paradigms that change people's ways of thinking on the subject.

Such revolutions in thought sometimes took time to develop, while others appeared relatively quickly. Since Gustavus Adolphus was the king of Sweden, he could force the adoption of the new paradigms and reorganization of the military necessary to make his changes happen relatively quickly. Conversely, though Rommel wrote a book called *The Attack* after World War I; it was only after the rapid fall of Poland that the world stood up and took notice of the blitzkrieg as refined by Guderian. Having high-level support for the new paradigm is essential to adapting quickly to each new age.

It is clear from the leading quote to Chapter 11 that Osama bin Laden understands that the rules have changed and that the media war is more important than ever. Al-Qaeda and its allies have applied the technologies of the Information Age more rapidly and effectively than the United States can currently counter [5].

12.1.2 The Information Age: What's New and What's Changed

The rules have changed, and we need to adapt. As we continue into the Information Age, a number of new technologies and techniques have appeared and more will arise in the future. The Internet, wireless communications, global media, nearly anonymous communications sources, personal digital assistants, digital manipulation, and many other technologies have appeared so quickly that their synergies have yet to be fully explored. During Operation Iraqi Freedom, for example, the U.S. military transmitted PSYOP messages on the Iraqi military communications channels [6], dropped leaflets on units that described how to surrender and the consequences of not following the directions [7], and sent e-mails directly to senior Iraqi personnel [8]. Alone, each of these techniques had reasonable success. However, these examples are just scratching the surface of what could be accomplished if one were to combine these and other Information Age technologies and techniques to defeat an enemy by primarily nonkinetic means. (Kinetic means still remain the best attention-getting, convincing, and coercive threat in the arsenal, but they do not win over minds and beliefs.) Moreover, succeeding in the Information Age requires success in the realm of influence operations, and extends both before and after a conflict.

As mentioned in previous chapters, the traditional line between military information operations and intelligence operations has become significantly blurred. The steps required in the military's intelligence preparation of the battlespace (IPB) looks a lot like traditional intelligence techniques. Advanced intelligence techniques that manipulate target information and information systems look a lot like traditional military IO.

This type of blurring has expanded to other realms as well. Military actions, even at the tactical level, have implications for strategic political and diplomatic efforts. State Department pronouncements can affect whether locals provide tips on or, alternatively, support paramilitary forces or terrorists. Influence operations can be successful in precluding an enemy attack, or when used by the enemy, can be useful in recruiting and motivating suicide bombers. Foreign national intelligence capabilities have been accused of being used for commercial advantage [9]. The Information Age has blurred the lines among all of the traditional instruments of national power—diplomatic, information, military, and economic. Success in the Information Age will require a greater contribution of closely coordinated capabilities currently "owned" by organizations such as the State Department, the intelligence community, and the Military Services, as well as other organizations.

A nation or non-nation group has different degrees of strength or depth in the four instruments of military power. When lacking in one area, they tend to exploit asymmetric advantages to counter their opponent's strengths. Al-Qaeda and a number of other anti-U.S. groups have achieved great success by trying to cause the United States to appear to be the bad guy, while the United States has sometimes played into their hands through events such as the Abu Ghraib prisoner abuse [10]. The old adage of "sticks and stones can break my bones, but words can never hurt me" is not a supportable claim in the Information Age. Words and thoughts shape beliefs, beliefs lead to actions, and the success or failure of those actions leads to the desire to acquire capabilities to take future actions. As mentioned earlier, the U.S. Secretary of Defense has recently stated that al-Qaeda is better at public relations than we are, getting its messages out much more quickly and effectively than the United States [5].

The revolutions in thought appearing in the Information Age are not just military revolutions; they are also political and social revolutions, revolutions in winning conflicts of the mind using the tools of information. The "New Battlefield" is primarily a battle between ideas, thoughts, and beliefs, rather than a battle of force. It reaches beyond the battlefield into the realm of how to prevent and deter conflicts, how to win without necessarily using kinetics, and how to win the postmilitary victory—winning the peace. A need for strong military capabilities remains, along with all of the other instruments of national power, but the key instrument of national power in the Information Age will be information. With information, battles can be precluded before they begin. Opponents can be dissuaded before they attack. Disaffected populations can be identified and their concerns addressed before they turn to violence. While there will still be violence and not every conflict can be defused, the ability to preclude even a subset of those that would otherwise have begun will be an improvement in future national capabilities. In contrast to the common U.S. view that any conflict is an aberration and is episodic, the information space requires a constant and sustained presence and level of activity to identify and defuse conflicts that we need not fight, and to shape in our favor those we do need to fight.

12.1.3 A Problem and an Opportunity

The problem is that we are still thinking at a national level as though we are incrementally improving our Industrial Age military. Our resource allocation continues to focus on improving conventional capabilities, with a few Information Age features added in. For example, the addition of a cell-phone detector on an EW aircraft is an Information Age feature added onto an Industrial Age platform [11]. This evolutionary approach would work if we were in an evolutionary situation, but we are actually in a revolutionary situation. We must take a moment to see what steps we need to take to succeed in the Information Age before we are hit by another Pearl Harbor or September 11.

The biggest stumbling block is the inertia of large organizations and the mindsets therein. The United States emerged from World War II as the least-devastated industrialized nation in the world. Because we won, and won decisively, there were two common beliefs that resulted: We did it right, and we will do it right the same way next time. The start of the Cold War and the use of proxies by the Soviet Union changed the rules. The U.S. military had to change from a focus on nuclear deterrence to Special Operations Forces and CIA-sponsored operations. Yet the focus of spending throughout the Cold War was on preparations for a potential World War III between the United States and the former U.S.S.R., even though, as stated by General Anthony Zinni in his farewell remarks, we could not come up with a plausible scenario in which the Soviets would start World War III [12]:

> The Cold War was ever present, and it was great for justifying programs, systems, and force structure—but no one seriously believed that it would actually happen. Still, it drove things. It drove the way we thought; it drove the way we organised and equipped; and it drove the way we developed our concepts of fighting.

So we continued to spend on nuclear offensive and defensive capabilities, as well as on a conventional military, until the Soviet Union collapsed and the United States had a very large national debt. The "peace dividend" was then vigorously sought, and the military was quickly drawn down. Then we found out that the world was even less stable than during the Cold War, so the military has been drawn back up again. However, many of the capabilities being added still emphasize success on the conventional battlefield. The 2006 QDR results demonstrate a continuation of the inertia behind that thinking. To quote Loren Thompson of the Lexington Institute [13], "Although the basic framework of threats with which the QDR began is still intact in the report, it was toned down by a year of deliberation and not a single signature weapon system has been terminated. That tells you that Rumsfeld's team is not so clear about what to do about this new environment."

This is not to imply that the U.S. military has not been innovative since the end of the Cold War. The United States won hands-down in Afghanistan against the Taliban, using new, adaptive combinations of conventional and unconventional capabilities. Having horseback-riding Special Forces personnel using satellite phones to call in conventional air strikes was not predefined as part of the table of organization and equipment, but a successful adaptation to the situation. In a similar manner, the Coalition military soundly and quickly defeated Saddam's military during Operation Iraqi Freedom, partly because of the expanded use of less conventional ways of doing things. Special Forces were used more extensively than ever before, operating both independently and in closer coordination with conventional forces. Better ways of doing things at the tactical level are appearing due to the skill and creativity of a number of our troops on the ground.

Thus adaptation is being demonstrated by the U.S. military, and there is an indication of recognition that we have to do something different in the Information Age. The 2006 QDR, however, did little to reallocate funds to improve our capabilities in the Information Age, or to develop the postconflict security and nation-building capabilities of the United States.[1]

While the actual implementation may take a number of years, and the nation's actual capabilities will change marginally over the first few years, planning now for the significant changes necessary to adapt to the Information Age will help us achieve this future goal. Rather than examining each budget year to decide what gets funded and what does not, a clear vision for the future and multiyear plan is necessary to achieve a revolution in national capabilities that can be developed and completed in the next 15 to 20 years.

We need to plan for a more significant change in the way we will operate in the Information Age across all instruments of national power. We currently have the Department of Defense to handle the military instrument of national power, the State Department to handle the diplomatic instrument, and the Commerce and Treasury Departments to regulate the economic instrument, but we have no cabinet-level office assigned to provide and wield the

---

1. Ted Bennett described in his popular article "The Pentagon's New Map" that there is a need for the United States to develop security, administration, and other nation-building capabilities as part of a combined interagency and military organization [14].

information instrument of national power. (See Figure 12.1.) While the intelligence community fulfills part of that role, there is much more that needs to be included, as described Section 12.2. This new organization will need to be built partially from scratch, but mostly by encompassing elements that already exist in other organizations but that are not yet operating with the cohesion of a central body.

Section 12.2 describes a vision for that future: to define the information sphere as its own domain similar to land, sea, air, and outer space, and to develop a national Military and Interagency Service to train, organize, and equip forces for use in that new domain.

12.2 The Information Sphere as a Domain[2]

The Information Age is upon us whether we like it or not, and the United States and other nations will need to adapt to adequately compete in this new

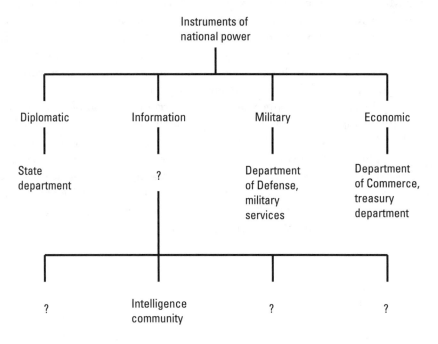

Figure 12.1 The information instrument of national power does not yet have representation at the highest levels of the government.

---

2. This section was developed in collaboration with Retired Lt. Col. U.S. Air Force Dennis Gilbert.

environment. As shown earlier, the nation states that first recognized the features of a new age and adapted to it quickly had an advantage over their competitors until those, too, adapted.

The maxim "train as you will fight" can be extended to "train, organize, and equip as you will fight." While it has been recognized for years that there are at least four instruments of national power—diplomatic, information, military, and economic—little has been done to develop an equivalent organization responsible for organizing, equipping, and training the information instrument of national power. Some capabilities to help succeed in the Information Age exist in portions of the Military Services, the intelligence agencies, and other interagency members such as the State Department, but no one has brought together all of these elements into a single, coherent framework to exploit the Information Age. And if the United States is not the first nation to do so, another nation or nonstate actor will be, which does not serve the United States well.

Part of the solution is to begin preparing for the creation of a new combined Military and Interagency Government Service to handle the information sphere as its own domain. Interest in whether or not, or how, to define information as its own domain similar to air, land, sea, and space has been previously discussed in Department of Defense circles [15, 16]. In this section, we propose a definition of a domain, list the features of a domain, and describe why the information sphere qualifies as a domain. Furthermore, we argue that not only does the information sphere meet all of the criteria of a domain, but it also has an additional important feature that distinguishes it from the other four physical domains. This distinction further supports our argument that the information sphere should be considered by the DoD as a separate domain. We then describe the types of benefits that could be obtained by treating the information sphere as a domain and recommend steps that the DoD would need to take to understand, resource, and be able to function effectively in this domain, thereby increasing the probability of dominating potential adversaries.

12.2.1 Definitions

In this stage of the Information Age, the U.S. military is grappling with whether or not it needs to define the information sphere as a domain, similar to the air, land, sea, and outer space domains that already exist. Some have spoken out for this definition, and some against it. A few notable people refer to this area as an "information domain," while others call it the "cyberspace domain," and still others refer to various elements of this area as either the "cognitive domain" or the "information environment" [16]. We believe that "information sphere" should be the preferred term, similar to the Russian term by the same name.

While considerable dialogue and research has been conducted on the subject, the author could not find an official military definition for a domain. Joint Publication 1-02, *The Department of Defense Dictionary of Military and Associated Terms*, does not define a domain. Therefore, we propose a definition of a domain, and what distinguishes one domain from any other domain, to help frame the discussion about whether to define the information sphere as its own domain.

12.2.1.1 Definition of a Domain

The *Webster's New Collegiate Dictionary* has two relevant general definitions of a domain:

1. A territory over which rule or control is exercised.
2. A sphere of activity, interest, or function.

Using these definitions as a guide, we propose the following definition for a domain for the DoD:

> *The sphere of interest and influence in or through which activities, functions, and operations are undertaken to accomplish missions and exercise control over an opponent in order to achieve desired effects.*

By breaking down this definition into its component parts, we can determine whether each of the existing four physical domains (air, land, sea, and space) qualify as a domain, and then determine whether the information sphere also qualifies as a domain. The key components of our proposed definition of a domain are that:

- It is a sphere of interest.
- It is a sphere of influence in that activities, functions, and operations can be undertaken in that sphere to accomplish missions.
- It is a sphere in which an opponent to friendly operations may function.
- It is a sphere in which control can be exercised over that opponent within or through that sphere.

Based on these components, it is clear that the four physical domains of air, land, sea, and space each qualify as a domain. Each has its own sphere of interest and sphere of influence. Aircraft fly missions, ships navigate the waterways (both surface and subsurface), land forces take and secure terrestrial objectives, and satellites orbit the Earth. In each physical domain an opponent can be present and can interfere with our operations. Moreover, the United States has military capabilities in each of these domains that can be used to control and dominate potential adversaries in each of these domains.

12.2.1.2 Proposed Definition for the Information Sphere's Domain

Current DoD Doctrine [17] defines the "information environment" as "the aggregate of individuals, organizations, or systems that collect, process, or disseminate information; also included is information itself." Regrettably, this definition puts the emphasis on the physical attributes of an information environment. In other publications, the information domain has been described as "the domain where information is created, manipulated, and shared," or "where information lives." The same authors [18] have defined the cognitive domain as the "domain of the mind of the warfighter and the warfighter's supporting populace." With this approach, the content, the connectivity, and the message have effectively been segregated. In this section, we argue that these preceding definitions diverge from the goals of the information operations mission area and the common understanding of strategic communication. Therefore, we first propose a definition of the information sphere, and second, a definition for the information sphere's domain.

The definition we propose for the *information sphere* is:

> *The space defined by relationships among actors, information, and information systems.*

To further elaborate on this definition, we also define actors, information, and information systems:

> *An actor may be a sender, a liaison, modifier, transferor, or recipient (either intended or unintended) of information. Information is the data and content being passed among actors via information systems. An information system is any communications, information storage, or information perceiving system, including couriers.*

Based on the preceding definitions of a domain and the information sphere, we propose a definition for the *information sphere's domain:*

> *The space of relationships among actors, information, and information systems that form a sphere of interest and influence in or through which information-related activities, functions, and operations are undertaken to accomplish missions and exercise control over an opponent in order to achieve desired effects.*

Note that the information itself is not the domain, nor is the domain simply the combination of information systems in which the information rides and resides. It is the *abstract space* defined by the *relationships* among actors, information, and information systems that define the information space and allow it to qualify as a domain.

Actors in this definition refer to people who originate the collection, use, modification or transfer of information. Therefore, people are the originators of the messages within information that are intended for human recipients and for information system recipients. The "influence" portion of the information domain focuses primarily on the thoughts and actions of the actors or humans that act or are acted upon within the information domain.

Information is the data being passed among actors via information systems. The data or content may or may not have meaning to the actor of an information system and may or may not be accepted by the actors of information systems. For example, the data may be encrypted, or it may be malicious software disguised as normal code. In either case, the raw information does not necessarily have meaning to the human or information system recipient, but does have meaning to the sender and the intended recipient, whether man or machine.

Information systems include all of the hardware and software necessary to store, retrieve, transmit, or collect information. Note that couriers are humans (or sometimes animals) used solely to carry messages, and therefore are an information system in this definition. Information systems may be altered by actors or the information they receive, and information systems can be used to alter both the information and the perceptions of the actors based on the information delivered.

The interactions among information, actors, and information systems define a complex space of relationships that defines the information sphere. However, the fact that the information sphere is not fully encompassed by any physical medium, but is a space of relationships, makes the information sphere distinct from other domains. The significance of this difference is defined further next.

We argue that the information sphere qualifies as a domain according to our preceding definition for the following reasons:

- The space of relationships among actors, information, and information systems forms a sphere of interest.
- It is a sphere of influence in that activities, functions, and operations can be undertaken in that sphere to accomplish missions.
- An opponent to friendly operations may function in that sphere.
- Control can be exercised over that opponent in or through that sphere.

In Section 12.2.2, we further elaborate on the features of a domain, in support of why we believe that the information space domain also qualifies as a domain.

12.2.2 Features of a Domain

In this section, we describe six features of a domain, and demonstrate how the information sphere satisfies these six features. Section 12.2.3 will describe what is different or unique about the information sphere that makes it unlike the air, land, sea, and outer space domains.

We argue that the six features of a domain are as follows:

- Unique capabilities are required to operate in that domain.
- A domain is not fully encompassed by any other domain.
- A shared presence of friendly and opposing capabilities is possible in the domain.
- Control can be exerted over the domain.
- A domain provides the opportunity for synergy with other domains.
- A domain provides the opportunity for asymmetric actions across domains.

We describe next how the four physical domains and the information sphere qualify as domains because they satisfy these six features.

12.2.2.1 Unique Capabilities Are Required to Operate in That Domain

Every domain requires unique capabilities in which to operate. For example, aircraft are required to operate in the air domain, spacecraft for the outer space domain, ships for the sea domain, and land systems for the land domain. Note that each of the unique capabilities can be readily differentiated from unique capabilities in other domains.

In a similar manner, information capabilities are required to operate in the information sphere. The information sphere requires unique equipment and personnel skills to function effectively, accomplish missions, and dominate any enemy presence in that domain. Information capabilities operating in the information sphere are both unique and distinct from the capabilities operating in the other domains.

As the information capabilities become more specialized to operate in the information sphere, the uniqueness and distinguishability of these capabilities will continue to grow. For example, the information sphere now has a set of unique equipment (materiel) and personnel skills required to effectively operate in and attempt to dominate the domain. With these new capabilities comes a range of unique support structures, such as doctrine, (including rules of engagement, tactics, techniques, and procedures, concepts of operations, and standard operating procedures), organization, training, materiel, leadership development, personnel, facilities, and policy. All of these support structures will tend to have some differences and similarities to support structures in the other domains.

12.2.2.2 A Domain Is Not Fully Encompassed by Any Other Domain

Another feature of a domain is that it is not fully encompassed by any other domain. For example, the air domain is not encompassed by the land domain, or vice versa. In a similar manner, the information sphere is not fully encompassed by the combination of land, sea, air, or outer space domains. The capabilities, missions, and dominance techniques of each domain remain unique and not provided by any of the other domains. For example, a tank does not have the capability to tap into a communication link, and a bomber does not normally have the ability to retrieve information from an enemy network. At the same time, such information sphere capabilities have been recently added to selected traditional platforms, such as a cell-phone monitor on an EW aircraft [11]. The information sphere has capabilities and functions that are meaningful only to the information sphere, such as routers and worms, and are not provided by assets developed for any other domain.

12.2.2.3 A Shared Presence Is Possible

Any domain can potentially be entered by opposing forces. This is not to say that every opponent is present in every domain, but that an opposing presence must be *possible* in order for the sphere of interest and influence to be considered a domain. For example, although there are few space-capable nations, the United States does share the outer-space domain with other nations' capabilities, which could oppose the United States capabilities in that domain. A potential shared presence is an essential feature of a domain, since dominance or control over the domain requires the possibility of an opposing presence or capability.

Note that until recently information systems rarely allowed for a shared simultaneous presence. Physical access to enemy information or the populace was precluded by distance and military capabilities. With the birth of the Information Age, however, the information sphere is frequently shared. Examples of this sharing include the range of information media, including the Internet, personal communications devices and digital assistants, local and wide-area networks, television, radio, print media, video and audio recordings, and other capabilities. All of these types of information systems are becoming even more interconnected as we continue into the Information Age. Due to the explosion of information and information capabilities, the information sphere allows for much more of a shared presence than ever before. As a result, dominance and control in that domain have become much more important than in the past.

Also note that the absence of a shared presence precludes defining a sphere of interest as a domain. For example, the field of military logistics could qualify for four of the features of a domain (the first and last two), but would not qualify as a domain because it cannot include an actual enemy presence. Therefore, there is no opponent over which to exert influence and control, and it should not be considered a domain. This important aspect of influence and exerting control distinguishes a domain from concepts that only define spheres of interest. The absence of an opponent and the opportunity to exert control over that opponent precludes many spheres of interest from being considered domains.

12.2.2.4 Control Can Be Exerted

The shared presence of a potential opponent in the sphere of interest generates the need to influence such opponents in the domain. By definition, an opponent opposes the realization of an effect desired by one side. Since a domain is a sphere of influence as well as of interest, then it must be possible for one side's influence in a domain to dominate the opposing side's influence in that domain. Therefore, a preponderance of influence of one side in a region of the domain defines control over that region by one of the two or more opposing sides.

For the information sphere, control can refer to the control of the information systems in a region, control of the access to the information in that space, or it may refer to the dominance of one belief over another in a region of the information sphere. As an example, air-to-air radars on fighter aircraft may try to jam or spoof the radars of opposing forces in the air domain, thereby attempting to control the information sphere, not the air domain. An example for the influence arena is the ability of one idea to dominate other ideas in the minds of a major population group.

12.2.2.5 A Domain Provides Opportunities for Synergy

The capabilities in a domain must be able to potentially provide synergistic opportunities with capabilities in other domains. The classic *Air-Land Battle Doctrine* was an excellent example of how the capabilities of the land and air domains could be mutually supportive. Airlift and sealift capabilities provide support to land forces, while space reconnaissance systems provide support to the other domains.

In a similar manner, the information sphere provides synergistic support to all the other domains, and vice versa. The ability to gather information directly from an enemy information source can assist air, land, sea, and space operations. Conversely, the ability to take out an enemy information system from the air can force the enemy to use an information system already compromised by our side.

The synergistic opportunities provided by information capabilities help define the information sphere as a domain. In a similar manner, the synergistic opportunities provided by capabilities in the other domains helps define the information sphere as a domain as well.

12.2.2.6 A Domain Provides Asymmetric Opportunities

Similar to synergistic opportunities are the opportunities for capabilities in a domain to gain an asymmetric advantage over opposing forces in other domains. For example, the Joint Fires Doctrine emphasizes the opportunity to use air assets as an asymmetric threat against opposing land and sea assets, while land or sea forces can be used to asymmetrically threaten enemy air assets. The principle of asymmetry must be a possibility for capabilities in a sphere of interest in order to be defined as a domain.

Information capabilities can provide an asymmetric threat against enemy capabilities in other domains. In his book, *The Next World War*, James Adams describes a case in which a computer virus is entered into a printer that is supposed to be delivered to an enemy air defense site. If the attempt had been successful, the ability for an information capability to neutralize an enemy air defense system would have been an example of an asymmetric opportunity [19].

12.2.3 Differences of the Information Sphere's Domain

Now that we have defined why the information sphere qualifies as a domain, this section will describe why the information sphere is unique compared to the air, land, sea, and outer-space domains. At the same time, we will describe why we believe that what makes the information sphere unique is yet one more reason why the information sphere should be treated as a domain by the Department of Defense.

Since the definition of the information sphere included actors, information, and information systems, then each of these three components must reside in a physical medium at any point in time. For example, an information server is either on the ground, underground, in the air, in space, on the sea, or under the sea. The server information system has a physical location that matters to the functioning of that system.

In a similar manner, the information itself is either being stored on an information system (including a courier), or is in some information conduit (including a portion of the electromagnetic spectrum) at any point in time. Finally, the human actors must be located in one of the four physical media. Figure 12.2 gives an example of how one might consider the information sphere with respect to the four physical domains.

Note that this figure shows that the information sphere is separate from each of the four physical domains, but is also accessible by, and provides access

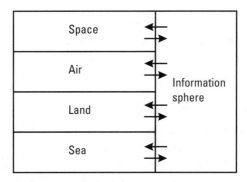

Figure 12.2 The information sphere is a unique type of domain, interacting with all four physical domains, but not encompassed by any or all together.

to, those domains. Information may enter or exit from a physical medium, but that may or may not be relevant or important. For example, if an intruder is seeking an entry point into a network, the physical location of the entry point may matter to the intruder. Once the intruder is on the network, however, then the physical location of the entry point and the location of any informational areas of interest are of less importance due to the degree of access provided by the network. What is important in this case is the relationship between security elements of the network (including people) and the intruder, and the relationship between the intruder and his objectives, rather than the physical location of the assets.

Note that the concept of entry and exit points from one domain to another is prevalent in all domains. Aircraft and spacecraft land on the ground or at sea. Ships dock at ports. Even among the four existing physical domains, there is a constant flow of entry and exit into or from each of the domains to the other domains. The same is true for the information sphere. There will always be entry and exit points from the information sphere to and from the other domains, as the purpose of most activity in the information sphere is to eventually affect something in the physical world. However, there are also actions and desired end states associated with operations *within* the information sphere that are unique to the information sphere, irrelevant to the physical space, and unaffected by the physical domain in which the information, actors, or information systems actually reside. For example, once one has full access to a computer network, it matters little if the perpetrator is in one physical domain (e.g., in an airplane) while the approved members of the network are in another (e.g., on land).

Each domain has actions that are dependent and actions that are independent of each of the other domains. Due to both the physical and abstract interactions and relationships between the actors, information, and information systems, however, the information sphere is not completely encompassed by any physical domains. For example, a distributed database that has elements either residing or in transit on land, in the air, on the sea, and in space is not contained or fully encompassed by any of the four media in which it is located or passes through.

Moreover, even the union of air, land, sea, and space media does not fully encompass the information sphere. The interactions we described for the information sphere often occur in a space of relationships, where the physical location of the actual components is irrelevant once access has been achieved. For example, the ability for two actors to interact in some way does not depend on the medium or media in which the information exchange occurs. What matters are the interaction and the relationship between the actors, information, and information systems. Moreover, shared presence within all four physical domains does not equate to dominance in the information sphere, either in the control of information access, information systems, or in the beliefs and perceptions of groups of interest within those four domains.

Furthermore, the desire to interact in some way is frequently independent of the physical locations associated with that desire to interact. The trust of the actors about the information they receive and their information systems relies less on the physical location and more on the security of the information systems and the reliability and track record of the actors and their information. The confidence one actor has in a particular piece of information has more to do with the track record of the information source and information system conduit and storage than in the physical location of any of the elements. Note that the desire to interact, trust, and confidence are all abstract concepts and relationships between actors that have a life of their own, independent of the physical location of the actors, information, or information systems.

It is these abstract relationships between actors, information, and information systems that define the interest and influence mechanisms in the information sphere. Since these relationships can be satisfied by a wide range of paths into, out of, and through various physical media, the value, benefit, and vulnerability of elements within the information sphere are relatively independent from the four physical domains.

In addition, the *desired effects* of an information activity usually reside eventually in one or more of the four physical domains. For example, the information activity may be to bring down an enemy air defense system, which opens the way for the air operations, which shapes the upcoming ground or sea battles. However, the phrase "resides eventually" means that there may be a significant delay between the initial information activity and any realization of effect in one or more physical domains. For example, the placement of a back door on a target server does not have an immediate effect other than the opportunity for access at a later date. Until that access is exploited, there is no physical manifestation of a desired effect beyond the ability to access.

As another example, competition between opposing thoughts or beliefs frequently has a delayed reaction. The concept of freedom, for example, is often dormant until the opportunity to be free, or to achieve increased freedoms, becomes available. In the conflict among beliefs, a thought that is planted may blossom many years after the fact after additional thoughts and physical events have occurred which support the originally planted thought.

Therefore, the fact that the actors, information systems, and information that comprise the information sphere must reside at any instant in one of the four physical domains is either secondary or irrelevant to the functioning of the relationships within the information sphere. Information easily transcends the barriers between the physical domains. The information sphere is a space in which the understanding of relationships in that space can lead to dominance over opponents in that space.

12.2.4 Benefits of the DoD Treating the Information Sphere as a Domain

This section presents some of the potential benefits of treating the information sphere as a domain equal to the already recognized domains of air, land, sea, and space.

First, in any conflict, there is always an advantage to the side that *understands and operates* within a domain better than the opponent does. This is true on land, air, sea, and outer space, and is also true in the information sphere. For example, the Joint Vision 2020 emphasis on information dominance and information superiority demonstrates the understanding that operating better than an opponent in the information sphere is a desirable objective. Obtaining information dominance in the information sphere is likely to lead to advantages and eventual dominance in the four physical domains as well via asymmetric effects. By defining the information sphere as a domain, a body of knowledge or military science of operating in the information sphere will soon be developed to improve understanding and consensus on the subject.

Second, *representing* the relationships of information among actors and information systems in a manner useful to planners and decision-makers will help improve the effectiveness and efficiency of operations in and through the information space sphere. For example, the ability to readily visualize relationships in a common format will facilitate a unity of effort and common understanding of objectives and constraints. Relevant and timely situation awareness, assessment, planning, and synchronization and deconfliction of actions and effects within and through the information sphere will be essential to success in that domain. Defining the information sphere as a domain should lead to an investigation and experimentation on a number of methods to represent these relationships, and the best-of-breed methods should emerge to enhance our capabilities in that domain.

Third, focusing and *preparing enhanced capabilities* in the information sphere will enable superior *influence and control* in the information sphere. Personnel with better skills and training will be more capable than their adversaries operating in the information sphere. The side with better equipment, doctrine, organizations, and leadership will have a significant advantage over the opposition in that domain. If the DoD chooses to define the information sphere as a domain, then the resourcing to effectively and efficiently function in that domain should follow soon after.

Fourth, defining the information sphere as a domain allows for increased emphasis on *planning and employing* all instruments of national power—diplomatic, informational, military, and economic—in a common, coordinated endeavor. Since information is a common element in the use of all instruments of national power, the ability to function effectively in the information sphere will encourage the coordination and synchronization of effects among all these instruments.

Fifth, defining the information sphere as a domain will help increase the emphasis on improved *information security*, which can and should lead to improved economic security and improved national security. The interconnected communications networks across the globe, and the heavy U.S. and Western reliance on these networks, has helped fuel a burgeoning economy, but at increased risk. Defining the information sphere as a domain will help define the common areas of interest between these sectors, and eventually lead to common or at least coordinated resourcing in the areas of information security.

Finally, defining the information sphere as a domain can help *focus national effort* on the important conflicts already ongoing in this domain. In addition to the skirmishes in cyberspace, the battle for the hearts and minds of many groups of actors worldwide has been raging since the birth of philosophies and political systems. For example, North Vietnam won the perception war even while losing the ground war at the time. This eventually led to winning the ground war after the United States had left South Vietnam. The current war on terrorism has highlighted the need to fight the battle for the minds in a coordinated fashion on a global scale on a consistent basis. In a battle of the minds, the *physical* location of the people believing in something is less important than the *dominance of that belief* over other competing beliefs, regardless of the physical domain. Although the United States may perceive itself to have an advantage in the cyberspace portion of the information sphere, it needs increased abilities in the influence portion of that domain associated with global, national, and ethnic opinion and attitudes towards the United States and its values. Defining the information sphere as a domain will help highlight the need for renewed effort and capabilities in this portion of the information sphere.

Information has always been used by, and important to, both military and civilian operations. What the Information Age has done is made the information sphere not only widespread but *shared*. There are few places in the world where the news media do not reach. Incidents in the remotest parts of the world often carry global implications beyond any time in previous history. The United States, its allies, and its opponents share the information sphere to a degree never before experienced in the history of mankind. As a result, conflict in the information sphere is becoming more prevalent and more important than even direct military action in many cases.

Once the information sphere is considered a domain, the pooling and sharing of information-focused personnel and materiel resources can be accomplished. New doctrine, organizations, and training will be developed, with the eventual goal of evolving a Military and Interagency Service to be the primary party responsible for dominating the information space sphere.

The information sphere will require its own Interagency and Military Service sooner than the outer-space domain because so much of the nation's capabilities and actions are already occurring within the information sphere.[3] Every Military Service is reliant on success and uninhibited secure operations in the information sphere. The information sphere is becoming increasingly shared, and not all actors sharing the space are friendly or cooperating. Moreover, the need to act decisively and to dominate *within* the information sphere is becoming more necessary today than even in the outer space domain.

12.2.5 How New Domains Are Formed

Since classical times, two domains of operation dominated military and civilian operations: land and sea. With the advent of powered flight in 1904, the road toward a third domain was begun. Actions by opposing elements in the air began during World War I. Over the years, the Army and Navy developed their own air capabilities, each with its unique concepts and capabilities. At the end of World War II, the Army Air Corps became the U.S. Air Force—almost 50 years after the first man-powered flight.

In a similar manner, the dawn of the Space Age in 1955 encouraged each of the U.S. Military Services to invest efforts in the space domain. By the mid- to late 1990s, the DoD defined outer space as a fourth warfighting domain, and a Unified Command for Space (US SPACECOM) was formed to combine these efforts. Even though outer space has been defined as a domain, it has not yet become its own Military Service, and for good reasons. The main reason is

---

3. While Martin Libiki argued for an information corps in *The Mesh and the Net*, our proposition is that the Information Service is interagency from the start, not just another Military Service [15].

that the U.S. military is still rightly focused on terrestrial operations. The focus on the outer-space domain is on how space can support each of the terrestrial Services, and less on interactions among assets in space. If and when mankind ventures more frequently and permanently into space, the need for a separate Space Service will probably become more compelling. For now, however, our needs are predominantly terrestrial, and the capabilities are primarily space-to-surface or surface-to-space rather than space-to-space.

Based on the preceding observations, the historical trends for recognizing new domains tend to follow this sequence:

1. The *capability* to begin to operate in that domain is developed, such as the first powered flight or the first space flight.
2. The capabilities to operate in that domain become relatively *commonplace*, such as air travel or shuttle launches.
3. The capabilities in that space *to affect capabilities* in the other domains become recognized and exploited.
4. Sufficient recognition of the *unique and synergistic* nature of capabilities in the domain are recognized and further developed.
5. *Institutional and financial support* for the domain is developed, which leads to all aspects of the DOTMLPF being developed by that new institution applying its resources.

In our opinion, the information sphere has already achieved the first four steps in this sequence. What remains is to recognize the need for an information sphere as its own domain and to plan for the creation, support, and evolution of the institution that will become the prime coordinator and lead agency for the information instrument of national power.

Such an endeavor is not a small undertaking. We estimate that it will probably take 15 to 20 years for such an institution to become fully operational. However, steps should be taken soon to ensure that this vision can be realized even within 20 years.

The creation of an Information Military Sphere Sevice will not and should not occur overnight. The DOTMLPF for an Information Sphere Interagency Service have not yet been thoroughly thought through and developed, nor has the necessary consensus been reached as to whether this is the direction to proceed. However, identifying the need and rationale for a separate information domain distinct from the physical air, land, sea, and space domains will help guide and frame the discussion of future benefits and requirements. Subsequent steps necessary to make the information sphere its own Interagency Service domain will follow.

Any new institution will need to compete for resources with existing institutions, and the fight for resources for a new Military Service will require significant effort. The DoD still has a structure in which the Services are encouraged to compete for resources. One of the first considerations should be how to reform that process so that the Information Sphere Interagency Service (ISIS) is less threatening to the existing organizations, both military and interagency. To quote General Zinni [12]:

> The National Security Act of 1947, for example, set up the most dysfunctional, worst organisational approach to military affairs I could possibly imagine. In a near-perfect example of the Law of Unintended Consequences, it created a situation in which the biggest rival of any U. S. armed service is not a foreign adversary but another one of its sister U.S. services. We teach our ensigns and second lieutenants to recognize that sister service as the enemy. It wants our money; it wants our force structure; it wants our recruits. So we rope ourselves into a system where we fight each other for money, programs, and weapon systems. We try to out-doctrine each other, by putting pedantic little anal apertures to work in doctrine centers, trying to find ways to ace out the other services and become the dominant service in some way.
>
> These people come to me and the other CinCs and ask, "What's more important to you—air power or ground power?" Incredible! Just think about it. My Uncle Guido is a plumber. If I went to him and asked, "What's more important to you—a wrench or a screwdriver," he'd think I'd lost my marbles. The real way this stuff gets worked out is not in the doctrine centers but out in the field. The joint commands and the component commanders can figure things out because we're the warfighters. We have to work things out, so we actually do ...
>
> We've had to be pushed into co-operating with each other by legislation. And those of us who have seen the light and actually put on joint "purple" uniforms—we've never been welcomed back to our parent services. We have become the Bad Guys. The only thing we are trusted to do is to take your sons and daughters to war and figure out ways to bring them back safely. Virulent inter-service rivalry still exists—and it's going to kill us if we don't find a better way to do business. Goldwater-Nichols is not the panacea everybody thinks it is.

It seems difficult to imagine that the national Information Sphere Interagency Service would be immune from such turf battles unless either the incentives to compete become reduced through improved legislation, or the ISIS is somehow budgeted independently from the existing Services and agencies. The former would be preferable, but the latter would be acceptable.

Or we could just stumble along, like General Braddock walking into the French and Indian ambush because he refused to recognize that he was fighting an enemy who was not going to fight fair—that is, fight according to the gentlemanly rules of engagement then used (and understood) by the British. Our enemies are not fighting fair right now; they are not, and have no intention of, fighting by our rules. We do not want to turn ourselves into our enemies, but instead of continuing to do the same things that we have always done and that no longer work, we need to adapt to the new age that we are in, and plan on how to succeed in the Information Age.

References

[1] Johnson, Spencer, *Who Moved My Cheese?* New York: Putnam Publishing Group, 1998.

[2] Christensen, Clayton M., *The Innovator's Dilemma*, New York: Harper Business Essentials, Harper Collins, 1997.

[3] Hart, B. H. Liddell, *Strategy*, New York: Praeger Paperbacks, 1954.

[4] "Basil Liddell Hart," http://en.wikipedia.org/wiki/B.H._Liddell_Hart, April 2005.

[5] Reuters, "Rumsfeld: Al Qaeda Has Better PR," CNN.com, February 17, 2006, http://www.cnn.com/2006/POLITICS/02/17/security.rumsfeld.reut/index.html.

[6] "Information Warfare Tools Rolled Out in Iraq," *Jane's Defence Weekly*, August 6, 2003.

[7] Friedman, Herbert A., "Operation Iraqi Freedom," Psywarrior Web site, http://www.psywarrior.com/OpnIraqiFreedom.html, October 2005.

[8] French, Matthew, "DOD Aims Psyops at Iraqi Officers," FCW.com, March 24, 2003, http://www.fcw.com/fcw/articles/2003/0371/web-psyops-03-21-03.asp.

[9] Chattergee, Pratap, "$pying for Uncle $am," *Covert Action Quarterly*, Winter 1996, http://mediafilter.org/caq/Caq55EconIntel.html.

[10] "New Abu Ghraib Abuse Images Revealed," MSNBC.com, February 15, 2006, http://www.msnbc.msn.com/id/11362767/.

[11] Fulghum, David A., "First Version of a New Electronic Attack System Goes to War," *Aviation Week & Space Technology*, November 7, 2005, p. 33, http://www.aviationnow.com/publication/awst/loggedin/AvnowStoryDisplay.do?pubKey=awst&issueDate=2005-11-07&story=xml/awst_xml/2005/11/07/AW_11_07_2005_p32-35-02.xml&headline=First+Version+of+a+New+Electronic+Attack+System+Goes+to+War.

[12] Zinni, General Anthony C., "Farewell Remarks at the U.S. Naval Institute," March 2000, http://www.rcaca.org/News-Zinni.htm.

[13] "Exclusive: What's in the QDR: Draft Boosts Spec Ops, UAVs, Cuts No Major Programs," *Defense News*, January 23, 2006.

[14] Barnett, Thomas P, M., *The Pentagon's New Map*, New York: G. P. Putnam's Sons, 2004.

[15] Libiki, Martin C., *The Mesh and the Net*, Institute for National Strategic Studies, National Defense University, McNair Paper 28, Washington D.C., March 1994.

[16] Alberts, David S., et al., "Understanding Information Age Warfare" *DoD Command and Control Research Program*, Washington, D.C., August 2001.

[17] Joint Publication 1-02, *Department of Defense Dictionary of Military and Associated Terms*, U.S. Government Printing Office, Washington D.C., April 12, 2001, p. 203.

[18] Alberts, David S., et al., *Network Centric Warfare*, 2nd ed., DoD C4ISR Cooperative Research Program (CRP) Publication, 1999.

[19] Adams, James, *The Next War*, New York: Simon and Schuster, 1998.

Appendix
IO Attack Desired Effects Definitions

These are definitions for the actionable near-term desired effects for the IO attack missions presented in Chapter 5. Tim Autry is the primary author of these revised definitions where they differ from the *Joint IO Handbook*.

The numbered definitions are from the Joint Forces Staff College, *Joint Information Operations Planning Handbook*, July 2003. Revisions to these definitions were created by Tim Autry and the author and forwarded to Air Combat Command as recommended updated definitions.

Degrade

1. "Damage done to the function is permanent, but only portions of the function were affected; that is, the function still operates, but not fully."
2. "A function's operation is permanently impaired, but the damage does not extend to all facets of the function's operation."

Comments: We do not view *degrade* as only permanent damage. Degrade is simply a reduction of a function's performance for *any* period of time. For example, we can degrade a radio signal by a given percentage (i.e., only partially audible). The function still exists, but its *performance* has been reduced. This effect ceases to exist the moment the jamming action is discontinued. Conversely, if instead of a jamming action we physically damage the radio antenna to some extent, the performance degradation is, in fact, permanent until the antenna is replaced or repaired.

Delay

1. As part of massing; that is applying a mutually supporting strategy to use different effects in rapid sequence to confuse or delay adversary response.
2. Cause adversary to waste combat power with inappropriate or delayed actions.

Comments: Note that the *IO Handbook* no longer includes delay as one of the original "five Ds" and defines delay as the cumulative (longer-term) effect on adversary decisions and actions. The near-term actions are defined as degrading information flow, sometimes by delaying the flow of information. The only change this author would like to make is to allow delay to also be applied as a near-term action for specific, identified information items critical to the enemy's decision process.

Deny

1. "Damage done to the function is only temporary, but all aspects of the function were affected."
2. "A function's operation is impaired over the short term, but the damage extends to all facets of the function's operation."

Comments: Deny versus *degrade* is more a difference in degree than duration. Both can be permanent or temporary; however, *degrade* is a reduction of performance, while *deny* is a complete removal of all or a portion of a function or information supporting a function, to the extent that the function can no longer be performed for a given period of time.

Destroy

1. "Damage done to the function is permanent, and all aspects of the function have been affected."
2. "A function's operation is permanently impaired, and the damage extends to all facets of the function's operation."

Comments: Agree all pertinent aspects of the function are permanently affected, or affected until damaged is repaired or element is replaced.

Disrupt

1. "Damage done to the function is temporary, and only portions of the function were affected."
2. A function's operation is impaired over the short term and the damage does not extend to all facets of the function's operation.

Comments: Like degrade and deny, we feel these definitions do not adequately distinguish these related, yet distinct, terms. We view the *disrupt* effect as an intermittent break in service (i.e., causing Internet service to a given entity to cut in and out). To clearly articulate the subtle differences between degrade, deny, and disrupt, we provide the following example: By causing a television picture to be "snowy" and its audio signal distorted, I have *degraded* it. Television service remains present, yet its performance has been affected. If I completely block the television signal (i.e., brute force jamming), I have completely *denied* the television service, for any period of time. Finally, by causing intermittent breaks in service (i.e., at times completely clear and at others nonexistent), I have *disrupted* the signal. The subtle differences between these three effects have even greater importance when determining what perception the planner wishes the enemy to have with relation to *why* the effect is occurring or *who* is causing it.

Exploit

1. Attempts to gather information that will enable opposition ability to conduct operations to induce other effects.

Comments: We agree with the inclusion of this effect and its definition.

Influence

1. "Selected projection or distortion of the truth to persuade the opposition to act in a manner detrimental to mission accomplishment while benefiting accomplishment of friendly objectives."
2. "To cause a change in the character, thought, or action of a particular entity."

Comments: We agree with the inclusion of this effect and its definition. We suggest adding "behavior" to the second definition. Note that "influence" is actually part of an influence operation, but as mentioned earlier, since any action can have an effect in the influence realm and vice versa, we included influence here as a desired effect within the IO attack desired effect definitions.

Manipulate/Control

The *IO Handbook* discusses control of the EW spectrum, control in support of IA, and command and control. It does not define control as a desired effect. However, the sample IO planning worksheet on page 82 lists examples of a pressure point being a factor that can be used to control behavior, while the sample shows "gaining political control over an area" as an action or theme to achieve interests.

Comments: In this author's opinion, manipulating data or routing access is a specific near-term action that is a distinct desired effect from the other desired effects described above. Furthermore, controlling access, controlling network behavior, and controlling larger organizational behavior are desired effects. Therefore, manipulate/control is a desired effect that can in some cases be achieved by near-term actions, and therefore should be included in the basic set of desired effects.

Mislead/Distract

1. Creation of a false perception that leads the opposition to act in a manner detrimental to mission accomplishment while benefiting accomplishment of friendly objectives.

Comments: We agree with the inclusion of this effect and its definition.

Penetrate/Access

1. The *IO Handbook* lists access or accessibility as a criterion for achieving other desired effects.

Comments: While it is true that access is a prerequisite for most subsequent IO actions, sometimes access is the desired effect in the near term to facilitate the necessary accessibility in the future. Therefore, the objective of some near-term desired effects is access, to set the stage for later activities. This is equivalent to capturing a key bridge even if the exploitation of that bridge will only occur later.

Stimulate

1. Stimulate is not listed in the *IO Handbook*.

Comments: One of the most common techniques used in EW is to stimulate the target to get it to react so the features of the targeted system and its procedures can be measured. In a similar manner, actions can be taken against information systems and networks to stimulate them to see how they behave in response to a specific stimulation. Stimulating a target to better understand the target, or to generate target insensitivity, is a near-term action with near-term desired effects.

Stop/Deter

1. Showstoppers;
2. Deter specific enemy behaviors.

Comments: The use of the term "stop" may be included in the definition of "deny" for some types of targets. However, deter usually applies to an organizational or national setting, and may require multiple actions to achieve a longer-term effect. However, at the tactical level, stopping a particular movement or transfer in its tracks may be the specific near-term desired effect that needs to be achieved.

Other Desired Effects

Note that inform and expose tend to be desired effects associated with influence operations and not with IO attack desired effects. Inform and expose are perfectly valid influence operation desired effects, and would appear on that list of desired effects. As mentioned earlier, influence is a desired effect within influence operations.

Protect is another desired effect, but usually associated with IO defense (such as CND, OPSEC, or military deception). The IO defense list of desired effects should include at least protect, deter, and influence.

List of Leading Quote References

Chapter 1:

Ball, Dr. Simon, "European History, 1500–2000," School of History and Archaeology, University of Glasgow, online course, http://www.arts.gla.ac.uk/History/Modern/Level1/appendix3.htm, November 2004.

Chapter 2:

Clarke, Richard, "Vulnerability: How Real Is the Threat?" interviewed on "Cyberwar," *Frontline,* April 24, 2003, © 2003 WGBH Educational Foundation. All rights reserved. http://www.pbs.org/wgbh/pages/frontline/shows/cyberwar/vulnerable/threat.html.

Chapter 3:

United States Marine Corps, Concept Paper, "A Concept for Information Operations," Marine Corps Combat Development Command, Quantico, Virginia, April 19, 2002.

Chapter 4:

Nixon, Richard M., "address before the National Association of Manufacturers, New York City, December 8, 1967. James J. Kilpatrick quoted a transcript in his syndicated column in *The Evening Star,* Washington, D.C., December 26, 1967, p. A13. Nixon's topic was the 'war in our cities.'" *Respectfully Quoted: A Dictionary of Quotations,* 1989, Number 1939.

Chapter 5:

Colmar, Wilhelm Leopold, Baron von der Goltz (1843–1916), http://www.firstworldwar.com/bio/goltz.htm.

Chapter 6:

Aeschylus, 525 B.C.–456 B.C., http://www.quotationspage.com/quote/28750.html; last accessed May, 2006.

Chapter 7:

Hartcup, Guy, *Camoflage: A History of Concealment and Deception in War,* Pennsylvania, PA: Encore Editions, 1980.

Chapter 8:

von Moltke, Helmuth, the Elder,

"Moltke's main thesis was that military strategy had to be understood as a system of options since only the beginning of a military operation was plannable. As a result, he considered the main task of military leaders to consist in the extensive preparation of all possible outcomes. His thesis can be summed up by two statements, one famous and one less so, translated into English as *No battle plan survives contact with the enemy and War is a matter of expedients.*" http://www.en.wikipedia.org/wiki/helmuth_von_Moltke_the_Elder.

Chapter 9:

DoDD 3600.1.Information Operations Policy, Washington, D.C.: Government Printing Office, December 2003.

Chapter 10:

Barrows, General (Ret.) R. H., Commandant, U.S. Marine Corps; King, Fleet Admiral E. J., 1942.

Chapter 11:

Osama bin Laden

Chapter 12:

Yogi Berra

About the Author

Patrick D. Allen is currently the senior lead systems engineer at General Dynamics Advanced Information Systems, is a certified Project Management Professional, and is a retired colonel in the Army Reserves. He is a director of the Military Operations Research Society (MORS), the cochair of the MORS IO working group, and a member of the PSYOP and OPSEC Communities of Knowledge and Practice (CKAPs).

Dr. Allen has designed and developed numerous real-world decision support tools and strategic technology plans for the Military Services, U.S. Pacific Command, U.S. Joint Forces Command, the Joint Special Operations Forces Institute, the Department of Defense, the Defense Advanced Research Project Agency, and the Intelligence Community. He is the award-winning designer and developer of major components of the Information Warfare Planning Capability, a fielded system that has been selected as the basis for the Department of Defense's next IO planning tool and the Air Force's Air Operation Center Strategy Development Tool. His work also includes the Integrated Battle Command Project, the Defense Against Cyber Attacks in Mobile Ad Hoc networks (DCA MANET), the Course of Action Process for computer network defense, IO test and evaluation support tools, the Interactive Intelligence planning and analysis tool, and many military models and simulations. Over his 25-year career, Dr. Allen has consulted internationally for Canada, the United Kingdom, and Sweden. He is widely published in the fields of operations research, and military modeling and simulation in analysis, training, testing, experimentation and planning. He has a B.S. in physics, an M.S. and Ph.D. in operations research, and a master's degree in strategic studies.

Index

4th Generation Warfare, 269
Able Danger, 248
Accusations, power of, 124–125, 195, 266–267
Acquisition
  planning, 62, 69–70, 224, 233–236
  testing, materiel, 224
Adaptive planning, 41
Adversarial reasoning technologies, 208–212
AFATDS, 192
AFDD 2–5, 18, 25, 57, 58, 111, 230
Afghanistan, 216–217
AFIT nation-building model, 202
AI programs, 183
Air Force IO doctrine, 17–18
Air Land Battle Doctrine, 296
Air Warfare Simulation. See AWSIM
Al Jazeera, 11–12, 124
Allegiances, shifting, 212–219, 270–272
ALPHATECH Inc., 198, 201
Al-Qaeda, 31–32, 80, 113, 114, 115, 170–171, 212, 215, 263, 269, 272, 286
Analyst Collaborative Environment (ACE), 193
Analyst's Notebook, 207
Army Field Manual on IO. See FM 100-6

Assessment, postexecution or postcrisis, 62, 67–68
Audit trails, 48, 87, 136–137, 195, 197, 199, 250
Austerlitz, battle of, 26–27
Automated response software, 66
AWSIM, 69, 235

BAE Systems, 198, 201
Baghdad Bob, 118, 124, 201, 273
Battle of ideas, 63
  See also Idea battlespace
Bayesian Belief Net, 202, 207, 211
Blue Capabilities Matrix, 73, 74, 77–79, 85, 86, 89, 90, 93, 97, 102, 199, 200
Benedict Arnold, 215
Bin Laden, Osama, 32–33, 212, 215, 232, 270, 278, 285
Blue Employment Option Selection, 74
Blue Force Tracking (BFT) technology, 159
Branches and sequels, 42, 46, 66, 67, 79, 90–91, 155–156, 166–167, 169, 178, 198, 263
  branches, 163–164
  computer network defense, 175–176
  monitoring, 161–166
  sequels, 162–163
Bribery, 212–215
Builder (Interactive Scenario Builder), 194

Cause-and-effect networks (CAENs), 73, 87–92, 95, 97–98, 101–102, 110, 129–130, 137, 157, 163–165, 198–200, 250
Calais deception operation, 136
Cause-and-effect chains, 37–38, 44–47
Cisco's Security Agent, 148
Civilian information infrastructure, 236
Civilian infrastructure defense, 63
Classification issues, 50–51
Clausewitz, 175
COAs, 85, 90, 105–107, 116, 190, 204, 211
　analysis, 74, 225
　audit trails, 106, 136–137
　comparison, Blue vs. Red, 91–97
　concept of operations, 92
　development, 187, 195–196
　development, PSYOP, 130
　employment options, 92, 97–100, 102
　employment/supporting, 92–95, 182, 196–198
　feedback from, 93
　in MILDEC, 135–140
　in OPSEC, 140, 143
　planning technologies, 195–204
　refinement of, 101–103
　supportability estimates, 92
　top-level/conceptual, 92–95, 101, 182, 196–198
COAP, 200
COAST, 183, 192, 194, 197–200
Collaborative Planning Tool (CPT), 192, 194
Collaborative Workflow Tool (CWT), 193
Combat camera, 223
Combat Evaluation Model (CEM), 69, 235
Command Post of the Future (CPOF), 159
Communications herding, 10
Communications monitoring, 11
COMPREHEND, 204
Computer network attack (CNA), 10, 73, 167
　in testing, 244–245
Computer network defense, 133–135, 171, 200, 226
　monitoring, 174–175
　planning, 146–153
　replanning, 175–176
Computer network exploitation, 73

Computer network operations (CNO), 16, 58, 60–61
Corps Battle Simulation (CBS), 69
Counter-intelligence, 133, 223
Crisis action planning, 62–64, 69
Critical Node Development, 227
Cyber conflict, 8–9, 301
　Palestinian-Israeli, 8–9, 33, 150
Cyber-attacks, 247
Cyber-defense, 247, 257
　civil, 236–237
　software, 152
Cyber-security, 236–237

DCA MANET, 149–50
Deception operations, historic, 26–27
Decision superiority, 4
Decision support, 66–67
Denial-of-service attacks, 9, 247
Desired effects. *See* Information Operations
DIME/PMESII, 18–20, 27–28, 79–80, 102, 185, 199, 200, 204, 208
Dispersion of forces, 27–28
Distributed control systems (DCS), 31
Distributed information systems, 28
Distributed planning, 49, 182–183
Distributed plans, 106–107
DLARS, 192
DoD IO Roadmap, 111, 181, 193
DoDD 3600, 18, 230
DoDD 3600.1, 60, 61
DoDD 3600.2, 111
Domain
　definitions of, 291–293
　features of, 294–297
　formation of, 302
DOTMLPF, 223, 224, 230–233, 237, 303

Effects
　direct, 87
　indirect, 87
　prediction technologies, 204–208
　synchronization, 157
　See also Information Operations
Effects-based operations (EBO), 18, 48, 200–202, 228–233
　USAF definition of, 18
Effects-based planning (EBP), 79, 82, 100, 200
Electromagnetic (EM) spectrum, 16

Electromagnetic signals, 249–250
Electronic
  combat operations, 111
  eavesdropping, 7, 104
  protection, 133, 140, 171
  tags, 6
  warfare (EW), 16, 58, 60, 61, 133, 167–168, 171, 223
Employment option analysis, 101
Enhanced Combat Assessment Tool (eCat), 192, 194
Enhanced Synchronization Matrix, 157–159, 194, 200
Enhanced Visualization (eViz), 194
Evidential reasoning tools, 185–186
Execution monitoring, 62, 66–67, 69, 155–179, 199–200
  attack, 166–169
  decision points, 163–164
  defensive IO, 171–175
  influence, 169–171
  MILDEC, 138–139
Execution Monitoring Tool, 91, 159–161, 165, 194, 200
Execution planning, 62, 64–65, 69
Exercises, 239–241
Experiment, 239–242
  limited objective, 242
Extend, 207

Feedback, 44–46, 67–68, 129, 155–157, 177–179
Feedback loops, 46, 62, 66
FM 100-6, 57, 58
FM 101-10-1, 58, 67, 178
FM 3-13, 58
Future Combat System, 149

GIANT (GPS Interference and Navigation Tool), 9
Grenada, invasion of, 228
Groups. *See* Messages
GPS jammers, countering, 9
Guadalcanal, invasion of, 228
Gustavus Adolphus, 284–285

Hackers, 8–9, 30–32, 150, 151, 174, 246, 284–285
Hactivism, 8
Hate messages, 259, 276
Hibernate, 104
honey pots, 45, 152
HUMINT, 143, 227

Idea battlespace, 47, 110, 112–113, 114–119, 126, 161, 169–171, 195, 199, 225, 257–259, 264–267
Inference relationships, 210–211
Influence arena, 47
Influence operations, 4, 13, 15, 38, 63, 109–131, 226, 228, 238, 2737
  change methods, 110
  doctrine, 110
  in Operation Iraqi Freedom, 41
  MOEs, 127–129
  See also Information Operations, influence
Influence, measuring, 127–129
Info Workspace (IWS), 103
INFOCON (information control), 148–149, 174–175
Information Assurance Test Tool, 246
Information,
  assurance, 17, 133
  CIA (confidentiality, integrity, availability), 4, 17
  conflict, 257–279
  definition of, 292–293
  inferiority, 4
  loss of confidence in, 76
  networks, dependence on, 29–33
  readiness, 20, 50, 224, 228–229
  security, 301
  sphere as domain, 283, 289–305
  sphere, definition of, 292–294
  superiority, 4, 59
  suppliers, 227
  system, definition of, 292–293
  warfare, DOD definition of, 5, 58
Information-in-warfare, 58
Information Support Server Environment (ISSE) Guard System, 51, 104
Information operations (IO)
  acquisition process, 224
  acquisitions planning, 233–236
  attack, 60
  attack options, 75
  capabilities, 16, 38–40
  capabilities, fragility of, 104, 232
  capabilities, offensive/defensive, 50

Information operations (IO) (continued)
  categories, 59–62
  classification issues, 50–51
  complexity, 47–49
  defend, 60
  defensive, 63, 134–135, 224, 229, 257
  defensive, planning, 133–153
  definitions, 5, 14, 17–18, 38–40, 43–44, 49, 51, 58, 60, 111
  desired effects, 73–75, 80, 86, 93, 299
  desired effects (actionable), 77, 78, 80
  desired effects, definitions, 309–313
  direct effects, 74, 75, 79
  doctrine, 57–62, 230
  facilities, 233
  fiscal planning, 233–236
  historic examples, 25–27
  implementation examples, 5–13
  indirect effects, 74
  influence, 60, 224. See also Influence operations
  information requirements, 49–50
  leadership development, 232
  legal aspects/issues, 224, 246–252
  logistics, 38, 223–252
  materiel, 231–232
  misperceptions of, 14–15, 17
  Navigator (ION), 193, 194
  offensive, 3,4, 43–44, 63, 224, 225, 257
  offensive, execution monitoring, 166–169
  operational functions, 16
  organization, 231
  personnel, 232–233
  physical scope, 42–43
  planning, 196–197. See also Planning
  planning characteristics, 37–52
  planning, information requirements, 225
  planning methodologies, 73–107
  planning space, 37–40, 100
  political constraints, 73. See also ROEs.
  time and space dimensions, 37, 40–44
  training, 231, 237–241. See also Training.
  undesired effects, 43–44, 86
Infrastructure Target System, 227
Integrated Battle Command (IBC) Project, 185, 199–200, 204, 208, 217

Integrated Information Infrastructure Defense System, 148
Intelligence preparation of the battlespace, 251, 285–286
Interactive Scenario Builder (Builder), 194
Internet
  conflicts, 8–9
  dependence on, 29–33, 29
  equalizer, 260
  terrorist use of, 30–31
  vulnerability of, 29–33
Internet Attack Simulator, 246
Intranets, military, 29
Intrusion detection system (IDS) software, 151–152
IO Joint Munitions Effectiveness Manual, 226, 229, 231–233
IO Roadmap, 60
ION (Information Operations Navigator), 193, 194
IOPC-J, 181, 182
ITEM, 69,235
iThink, 207
IWPC, 181–183, 188, 191–195, 197, 200
  COAST, 183, 192, 194, 197–200
  Collaborative Planning Tool (CPT), 192, 194
  Collaborative Workflow Tool (CWT), 193
  ION (Information Operations Navigator), 193, 194
  PSYOP planning module, 192

Java Causal Analysis Tool (JCAT), 207
JIAPC, 48, 227, 231, 233
Joint Expeditionary Force Experiment '04, 190, 194
Joint Fires Doctrine, 297
Joint Futures Battle Lab, 19–20, 230
Joint Information Operation Warfare Center (JIOWC), 193, 231
Joint Information Operations (IO) Planning Handbook, 57, 74, 309–312
Joint Munitions Effectiveness Manual (JMEM), 65, 178, 226, 229
Joint Officer's Staff Guide 2000, 186
Joint Simulation System. See JSIMS
Joint Vision 2010, 59
Joint Vision 2020, 300
Joint Warfare Analysis Center, 227

Joint Warfighting Simulation. *See* JWARS
JOPES, 186, 195, 204
JP 1-02, 291
JP 2-01, 227
JP 3-13, 5, 16, 25, 57–58, 60, 74, 109, 111, 133, 230
JP 3-53, 112, 116, 127
JSIMS, , 69, 235
JTF-CND, 226
JTT, 192
JWARS, 69, 235
JWICS, 29

Kobayakawa, Hideaki, defection of, 213, 215
Kosovo campaign examples, 6–8, 50, 82, 104
Kurdish autonomy example, 217–218

Land Information Warfare Agency (LIWA), 7
Legal issues, 246–252
Lenin, Vladimir Ilyich, 27
LIVE, 204
LPD (low probability of detection) attacks, 151

Management by exception, 84
Marj Dabiq, battle of, 213
Markov and Hidden Markov Models, 207–208
Materiel acquisition testing, 224
Matrix, comparison, 96
Maverick, 246
Measures of causal linkage (MOCs), 128–129, 135, 170
Measures of effectiveness (MOEs), 84, 90, 128–129, 135, 170, 250
    influence operations, 127–129
    message delivery, 128–129
    messages, 110
Measures of performance (MOPs), 127–129, 135, 170
Media, Western, 119–120
    opponents' exploitation of, 11–12, 170
Messages, 47–48, 169, 171
    attacks upon, 262–263
    attention, 110, 114, 119–120, 122, 169
    change methods, 120–127, 130
    conflicts/consistency among, 118

    co-opting, 122–123
    creating, 121
    cultural, 12–13
    delivery mechanisms, 110, 127, 130
    delivery MOEs, 127–129
    delivery MOCs, 110, 128–129
    discrediting, 123–124
    distracting from, 125
    group interactions, 118–119, 226
    groups and, 112, 129, 199, 225, 228, 258
    groups of interest, 47, 114–120
    hate, 259, 276
    identity, 258–260
    indicators, 109
    interaction among, 259
    methods, 110
    modifying, 122
    MOEs, 110, 130
    MOPs, 130
    orthogonality, 118, 262
    planning, 260
    PSYOP planning, 261, 262
    public affairs planning, 261. *See also* Public affairs.
    purposes of, 110, 112–114, 199
    sensitivity to, 262
    subverting, 123–125
Military deception (MILDEC), 111, 112, 133–140, 171, 199
    COAs in, 135–140
    execution monitoring, 138–139
    monitoring, 171–174
    planning, 135–140, 144–146
Military Decision Making Process (MDMP), 186, 195
Military Intelligence Data Base (MIDB), 87, 198
Military Standard 25–25, 163–164
Military testing, 15. See also Testing.
Milosevic, Slobodan, 7–8, 42
MOEs. *See* Measures of effectiveness
MOCs. *See* Measures of causal linkage
MOPs. *See* Measures of performance
Modeling and simulation tools, 182, 205–207

Napoleon, deception operations of, 26–27
Nation building, 41, 182
    Afghanistan, 216
    AFIT model, 202

National Intelligence Support Team, 227
Network combat operations, 111
Network Exploitation Test Tool, 246
NGOs, 201, 204
NIPRNET, 29
Nonadversarial reasoning technologies, 208–209
Nonadversarial reasoning, 212–219

Observation points, 40, 42
Operation Desert Storm examples, 6, 90, 93–95, 104, 128–129, 136, 138, 162, 165, 167–168, 204–205, 273
OODA loops, 14, 102, 219
Open-source information, 7, 227
Operation Earnest Will example, 81–82
Operation Iraqi Freedom examples, 9–10, 41, 159, 202, 285, 288
Operation Just Cause example, 205
Operational Assessment Tool (OAT), 202
Operational net assessment (ONA), 19, 200, 229
Operations planning, 62–69
Operations security (OPSEC), 111
OPNET, 148
OPSEC (operations security planning), 111, 133, 134, 140–146, 199
  COAs in, 140, 143
  plan monitoring, 171–174
Oracle, 104, 194

Palestinian-Israeli cyber conflict, 8–9, 33, 150
Passing electron analogy, 249
Phases of operation, 41
Physical security, 61, 133
Planning, 196–197
  acquisition, 62, 69–70, 224
  adaptive, 62, 65
  automated, 183–184
  crisis action, 62–64, 69
  defensive information operations, 133–153
  deliberate, 62–63, 69
  execution monitoring. See Execution monitoring
  execution, 62, 64–65, 69
  feedback in, 177–179
  influence, 109–131
  methodologies, 73–107
  operations, 62–69
  persistent, 62, 65–66
  process workflows, 186–189
  replanning, 156, 175–176, 200
  scalability, 100–103, 200
  semiautomated, 182–195
  technologies, COA, 195–204
  technologies, defensive information operations, 133–153
  tools, 181–219
  validated planning factors, 156, 177–178
  visualization, 184–185
PMESII. *See* DIME/PMESII
Political constraints. *See* ROEs, political constraints
Postcrisis or postexecution assessment, 62, 67–68
Prediction technologies, 204–208
Predictive battlespace awareness (PBA), 20
Propaganda
  Iraqi examples, 12, 118, 124, 168, 201, 273
  of the deed, 122
  Serbian examples, 6
Protection issues, 50–51
PSYOP, 109–131
  ACTD, 127
  COA development, 130
  in training, 238
  task development wizard, 199
Public affairs, 109, 111–112, 265
  constraints on, 126–127
  planning, 199, 261

Rainbow plans (WWII), 62
RAND Corporation, 228
Reachback centers, 90, 91, 103, 195
Reachback support, 38, 48, 68, 74, 227
Redundant links (for infrastructure), 63
Remote communications, 28
Request for detailed planning (RFDP), 102, 193, 195, 199
Requests for analysis (RFAs), 48, 102–103, 130, 192–193, 199
Requests for information (RFIs), 48, 102–103, 158
ROEs (rules of engagement), 78, 81–82, 87, 90, 250–251
  audit trail, 87

guidelines and constraints, 83–85, 88, 98–100, 197–198, 200
political constraints, 79–87

SCADA systems, hacking, 31–32
Scalability. See Planning
SEAS, 202–204
Security, multilevel, 104–107
Sekigahara, battle of, 213
Sequels. See Branches and sequels
Shifting allegiances, 212–219, 270–272
SIGINT, 134, 143
SIPRNET, 29
Skybox Security, 148
Social engineering, 16
SOFTools, 188, 190–191
Special Technical Operations (STO), 15
Stability operations, 41, 182
Standing Joint Force Headquarters, 229
Strategy Development Tool (SDT), 198, 201–202, 207, 211–212
Strategy-to-task, 88, 89
Sun Tzu, 25–26, 135–136, 217
Sunni insurgency, 217
Support infrastructure, military, 28–29
Symantec Internet Security, 148, 152
Synchronization
   analysis, 225
   effects, 157
   matrix, 157–159
   matrix, enhanced, 157–159
   observation points, 157–158
System of systems analysis (SOSA), 229
Systems dynamics models, 202

TacWar, 69, 235
Taliban, 212, 262, 288
Tannenberg, battle of, 50
Target considerations, 86, 87
Target decoys, 6
Target Prioritization Tool (TPT), 194
TBONE, 192
TEL-SCOPE, 194, 200
Testing
   archiving results, 246
   classification constraints, 238, 243
   computer network attack, 244–245
   developmental, 242–243
   information operations in, 242–246
   operational, 242, 244
   test and evaluation master plans (TEMPS), 242
THUNDER, 69, 235
Title 10, 224, 251–252
Title 50, 224, 251–252
Tobacco industry, 123–124
Tora Bora, 212, 215, 270
Training, 231, 237–241
   train as you fight, 231, 237, 238, 290
   exercises, 224
   computer network attack representation, 239
   computer network defense representation, 239
   computer network operations representation, 239–241
   OPSEC in, 238
   peacetime, 51
   PSYOP in, 238
   role players, 238
Transformation experiments, 224
Trojan Horse, 16, 40
Trusted Network Environment, 51, 104

UN Manual on the Prevention and Control of Computer-Related Crime, 249
Undesired effects. See Information Operations
USAF Predictive Battlespace Awareness Project, 207
USJFCOM J9, 235

Validated planning factors, 156, 177–178

WEEMC, 192

XML Briefing Composer (XBC), 194

Zombie software, 150

The Artech House Information Warfare Library

Electronic Intelligence: The Analysis of Radar Signals, Second Edition, Richard G. Wiley

Electronic Warfare for the Digitized Battlefield, Michael R. Frater and Michael Ryan

Electronic Warfare in the Information Age, D. Curtis Schleher

Electronic Warfare Target Location Methods, Richard A. Poisel

EW 101: A First Course in Electronic Warfare, David Adamy

Information Operations Planning, Patrick D. Allen

Information Warfare Principles and Operations, Edward Waltz

Introduction to Communication Electronic Warfare Systems, Richard A. Poisel

Knowledge Management in the Intelligence Enterprise, Edward Waltz

Mathematical Techniques in Multisensor Data Fusion, Second Edition, David L. Hall and Sonya A. H. McMullen

Modern Communications Jamming Principles and Techniques, Richard A. Poisel

Principles of Data Fusion Automation, Richard T. Antony

Tactical Communications for the Digitized Battlefield, Michael Ryan and Michael R. Frater

Target Acquisition in Communication Electronic Warfare Systems, Richard A. Poisel

For further information on these and other Artech House titles, including previously considered out-of-print books now available through our In-Print-Forever® (IPF®) program, contact:

Artech House
685 Canton Street
Norwood, MA 02062
Phone: 781-769-9750
Fax: 781-769-6334
e-mail: artech@artechhouse.com

Artech House
46 Gillingham Street
London SW1V 1AH UK
Phone: +44 (0)20-7596-8750
Fax: +44 (0)20-7630-0166
e-mail: artech-uk@artechhouse.com

Find us on the World Wide Web at: www.artechhouse.com